T0281460

Textile Science and Clothing Technology

Series editor

Subramanian Senthilkannan Muthu, Kowloon, Hong Kong

More information about this series at http://www.springer.com/series/13111

Miguel Angel Gardetti
Subramanian Senthilkannan Muthu
Editors

Organic Cotton

Is it a Sustainable Solution?

 Springer

Editors
Miguel Angel Gardetti
Center for Studies on Sustainable Luxury
Buenos Aires, Argentina

Subramanian Senthilkannan Muthu
Head of Sustainability
SgT & API
Hong Kong, Hong Kong

ISSN 2197-9863   ISSN 2197-9871 (electronic)
Textile Science and Clothing Technology
ISBN 978-981-13-4238-7   ISBN 978-981-10-8782-0 (eBook)
https://doi.org/10.1007/978-981-10-8782-0

This Springer imprint is published by the registered company Springer Nature Singapore Pte Ltd.
The registered company address is: 152 Beach Road, #21-01/04 Gateway East, Singapore 189721, Singapore

Preface

The purpose of this book is to make a contribution to the discussion with various specialists on whether organic cotton is sustainable or not. This is not about drawing conclusions but rather providing data to cast light into this discussion.

The book begins with a paper by Ali Serkan Soydan, Arzu Yavas, Gizem Karakan Günaydin, Sema Palamutcu, Ozan Avinc, M. Niyazi Kıvılcım, and Mehmet Demirtaş titled "Colorimetric and Hydrophilicity Properties of White and Naturally Colored Organic Cotton Fibers Before and After Pretreatment Processes". This chapter researches colorimetric (CIE L^*, a^*, b^*, C^*, h°, K/S, whiteness properties, etc.) and hydrophilicity properties of two white (Nazilli 84 S and Aydın 110) and three naturally colored (Emirel, Akdemir, Nazilli DT-15) organic cotton fiber types under review, before and after scouring (with NaOH), conventional bleaching (with H_2O_2), and the combined application of scouring and bleaching (scouring + bleaching) processes in comparison with their greige (untreated) counterparts.

The next chapter, "Physical Properties of Different Turkish Organic Cotton Fiber Types Depending on the Cultivation Area" was written by Sema Palamutcu, Ali Serkan Soydan, Ozan Avinc, Gizem Karakan Günaydin, Arzu Yavas, M. Niyazi Kıvılcım, and Mehmet Demirtaş. The measured and recorded data are analyzed with the Least Squares Fit model statistical evaluation method to accomplish Analysis of Variance and Effect Tests. Statistical evaluation has been designed to evaluate the influence of dependent variables of fiber type, location, and year in the independent fiber properties of length, strength, and fineness (micronaire).

Following "Sustainability Goes Far Beyond "Organic Cotton." Analysis of Six Signature Clothing Brands" was developed by María Lourdes Delgado Luque and Miguel Angel Gardetti. This chapter analyzes five Spanish signature fashion brands based on the sustainability criteria defined by the authors. For such purpose, all the public information referred to by the brands: websites, newsletters, articles, references from organizations, and case studies, if any, is studied. Each of the designers or owners of these microenterprises are also interviewed. All of this is compared to a model developed by the authors that addresses the meaning of being sustainable in the textile and fashion world.

Moving on, Gizem Karakan Günaydin, Ozan Avinc, Sema Palamutcu, Arzu Yavas and Ali Serkan Soydan developed "Naturally Colored Organic Cotton and Naturally Colored Cotton Fiber Production". White cotton fiber is one of the most chemically intensive crops cultivated. Though grown on 3–5% of the world's farmland, it is liable for the usage of 25% of the world's pesticides. For these aforementioned reasons, organically grown naturally colored cotton fiber has attracted a massive attention over the past few years. This chapter describes in detail a comprehensive review of naturally colored organic cotton fibers, naturally colored cotton fiber types, their properties, their production and their recent developments from a broad perspective and from many different angles.

The chapter called "Organic Cotton and Cotton Fiber Production in Turkey, Recent Developments" was written by Gizem Karakan Günaydin, Arzu Yavas, Ozan Avinc, Ali Serkan Soydan, Sema Palamutcu, M. Koray Şimşek, Halil Dündar, Mehmet Demirtaş, Nazife Özkan, and M. Niyazi Kıvılcım. This chapter deals with organic and conventionally grown cotton fibers with a broad perspective in terms of cotton fiber cultivation and recent development about these fiber types in Turkey. First, details are provided about organic cotton and organic cotton fiber cultivation in Turkey, organic cotton growing regions in Turkey, limitations for the organic cotton markets, lack of information on cost of production, marketing and future trends. Moreover, information about general cultivation in lands and cotton fiber yield in Turkey is given in detail, as well as information about the diseases and pests encountered during the cotton fiber cultivation.

In turn, in their paper "Organic Cotton and Its Environmental Impacts" P. Senthil Kumar and P. R. Yaashikaa investigate that the organic production is not really any more or any less ecologically well disposed than current ordinary cotton generation. For the textile procurer, there is no contrast between routinely developed cotton and organically developed cotton as to pesticide build-ups. Developing natural cotton is more demanding and costly than developing cotton routinely. Organic generation can be a challenge if bug weights are high; however, with work and experience, it could give premium value to cultivators willing to address these difficulties.

The next chapter, "Organic Cotton Versus Recycled Cotton Versus Sustainable Cotton" was developed by P. Senthil Kumar, and A. Saravanan. Organic cotton is cotton that has been developed without manures and pesticides, with advance biodiversity, organic cycles, and soil health. In contrast, natural cotton makes cotton development "cleaner," giving both natural and ordinary cotton experience a similar assembling process, which is water and vitality concentrated. Recycled cotton is repurposed, post-modern or post-shopper cotton that would somehow or another be considered straight up: squander for the landfill. The pieces of such cut and sew jobs are post-mechanical cotton "squander" with the ability of being reused. Contingent upon how reused cotton is utilized, it can possibly extraordinarily decrease water and vitality utilization in reasonable design and attire, and diminish landfill waste and space. Cotton development is related to various social, financial, and natural shortcomings that weaken the piece sustainability.

Completing the book, Seyda Eyupoglu prepared a chapter titled "Organic Cotton and Environmental Impacts". This chapter investigates organic agriculture, organic cotton agriculture, comparison of conventional cotton agriculture with organic cotton agriculture, environmental impacts of organic cotton agriculture, and use of organic cotton products. And the final chapter contains conclusions and recommendations.

It is important to highlight that all of these diverse contributions represent a great step forward in expanding the insights in this field. It is certainly the most comprehensive collection of writings on this subject area to date. Note that this initiative has received a wide international response, and it is expected to continue stimulating further debate.

Buenos Aires, Argentina   Miguel Angel Gardetti
Hong Kong, Hong Kong   Subramanian Senthilkannan Muthu

Contents

Colorimetric and Hydrophilicity Properties of White and Naturally Colored Organic Cotton Fibers Before and After Pretreatment Processes

Ali Serkan Soydan, Arzu Yavas, Gizem Karakan Günaydin, Sema Palamutcu, Ozan Avinc, M. Niyazi Kıvılcım and Mehmet Demirtaş

Abstract It is widely known that conventionally grown cotton fiber/fabrics/apparel has chemical residues on the cotton which may cause cancer and some other health related troubles. It is also certain that organic cotton production does not consume most synthetically compounded chemicals (fertilizers, insecticides, herbicides, growth regulators and defoliants) which are suggested for only conventional cotton production. Therefore, organic cotton production lead to much more environmentally cotton fiber production in comparison to conventional cotton fiber growing. So, in this chapter, colorimetric (CIE L^*, a^*, b^*, C^*, $h°$, K/S, and whiteness properties etc.) and hydrophilicity properties of studied two white (Nazilli 84 S and Aydın 110) and three naturally colored (Emirel, Akdemir, Nazilli DT-15) organic cotton fiber types was investigated before and after scouring (with NaOH), conventional bleaching (with H_2O_2) and the combination application of scouring and bleaching (scouring + bleaching) processes in comparison with their greige (un-treated) counterparts. Greige (un-treated) Akdemir naturally colored organic cotton fiber displayed the reddest (with the highest a^* value), the yellowest (with the highest b^* value) appearance, the highest chroma (the most saturated), the lowest lightness (the darkest) and the highest color strength (the strongest color yield) and therefore the strongest color shade amongst the studied greige (un-treated) naturally colored organic cotton fibers. After scouring process, all three naturally colored organic cotton fibers congruously exhibited darker [with the lower lightness (L^*) values and higher color strength (K/S) values], slightly redder (slightly higher a^* values) and slightly less yellow (slightly lower b^* values) appearance in comparison to their greige (un-treated) counterparts. Overall, it can be concluded that solely bleaching

A. S. Soydan · A. Yavas · S. Palamutcu · O. Avinc (✉)
Textiles Engineering Department, Pamukkale University, Denizli 20016, Turkey
e-mail: oavinc@pau.edu.tr

G. K. Günaydin
Buldan Vocational School, Pamukkale University, Buldan, Denizli, Turkey

M. Niyazi Kıvılcım · M. Demirtaş
Cotton Research Institute, Nazilli, Aydın, Turkey

© Springer Nature Singapore Pte Ltd. 2019   1
M. A. Gardetti and S. S. Muthu (eds.), *Organic Cotton*, Textile Science
and Clothing Technology, https://doi.org/10.1007/978-981-10-8782-0_1

process (without any prior scouring process) and combination sequential usage of scouring and bleaching processes (scouring then bleaching = scouring + bleaching) generally did not significantly affect the color properties of studied naturally colored organic cotton fibers leading to similar close colorimetric performance with their greige (un-treated) counterparts. So, after the bleaching process, scoured naturally colored organic cotton fibers which darkened due to the scouring process roughly turned back to their original colorimetric levels of greige (un-treated) versions. In this case, if the naturally colored organic cotton fibers are blended with the normal white and off-white organic cotton fibers or other cellulosic fibers, applied bleaching process does not cause a significant color change in the naturally colored organic cotton fibers and this indicates that they will approximately remain at the same color property levels as their greige (un-treated) counterparts. Moreover, the bleaching process following the scouring process slightly increases the hydrophilicity values of both white and naturally colored organic cotton fibers leading to more hydrophilic fibers.

Keywords Organic cotton · Naturally colored cotton · Color · Whiteness Hydrophilicity · Pretreatment · Scouring · Bleaching

1 Introduction

Organic cotton fiber production does not consume most synthetically compounded chemicals (fertilizers, insecticides, herbicides, growth regulators and defoliants) which are suggested for only conventional cotton production leading to more sustainable and ecological way of production. In Turkey, not only the white and off-white but also naturally colored organic cotton fibers are produced. Naturally colored cotton has a long history and the cultivation history of naturally colored cotton dates back approximately 5000 years [1–3]. For example, it is known that naturally colored cottons were cultivated and utilized in South and Central America around 2300 B.C. [4]. Even though naturally colored cotton exhibit a long cultivation history, the cultivation of the naturally colored cotton plant nearly ceased for a long period of time. Since, naturally colored cotton fibers are generally regarded as inferior to common white cotton owing to their lower yield, their shorter, and weaker fiber types and problems regarding the repeatability and shade variability of their color [4]. Indeed, it was stated that brown and green naturally colored cotton fiber display approximately 33.6 and 41.9% lower yields than common white cotton fibers, respectively [5, 6]. It is also reported that naturally colored cottons possess lower boll numbers, lower boll mass and also lower lint yield. What is more, up to 17.4% fiber length reduction, lower micronaire index and lower strength but higher elongation was measured in the case of colored cotton in comparison to common white cotton [5, 6]. Naturally colored cotton fibers could be too short and weak to be spun into finer yarn counts and moreover there is a constraint of non-existence of petitive different colors and shades [7].

Naturally colored cottons, exhibiting a very small niche market, available today are generally shorter, weaker, and finer than regular Upland cotton fibers, nonetheless these fibers can be spun effectively into ring and rotor yarns for many different end uses [8]. However, the cultivation of naturally colored cotton has captured the attention lately due to the increasing environmental concerns [9]. Although the yield of naturally colored cotton is lower than that of common white cotton, naturally colored cotton sometimes can be sold at three to four times more expensive prices [10]. Even though naturally colored cotton fiber is generally marketed at higher prices than common white-colored cotton, the cost of their wet processing is significantly lower than that of common white cotton. Since, dyeing process is not needed and dyes and their related auxiliaries are generally not used for naturally colored cotton fibers because of their natural inherent color characteristics leading to more environmentally friendly and cheaper textile processing due to their lower water, chemical and energy consumption and therefore lower wastewater load. Moreover, it was stated that naturally colored cotton fibers do not exhibit fading even after many washings [11]. It is conceived to be disease resistant during its growth [9]. It was also reported that naturally colored cotton fiber is more resistant to drought, so it can be easily grown in arid regions without loss of productivity. At the same time, the naturally color cotton fiber has been found to be resistant to many pests and diseases threatening common white cotton fiber [10]. Naturally colored cotton fiber also displays flame resistance performance and therefore higher limited oxygen index value owing to the presence of more heavy metals than common white cotton fiber [12]. Furthermore, the natural pigments of the naturally colored cotton fiber procure protection from UV rays of the sun for the embryonic cotton seed and also these fibers procure very good sun protection properties with high UPF values due to their colored pigment presence [13, 14]. Due to these positive attributes of naturally colored cotton fibers, cotton fiber breeders are working to improve the properties and quality of naturally colored cotton by creation of their hybrids via selective breeding leading to the production of better and more productive superior varieties of naturally colored cotton [4, 15, 16]. For instance, the fiber lengths of naturally colored cotton are shorter, cotton plant breeding researches continues to increase the fiber length. Moreover, cotton breeders are continuing their efforts to expand the number of naturally colored cotton colors and to improve fiber quality and yield.

Naturally colored cotton is genetically inherited characteristic that occurs as result of the of pigment presence in the cellulose of the cotton fiber. Although the natural color cotton has many different colors, the most common ones are the ones in different shades of green and brown [17]. Indeed, sales of different shades of brown and green organic naturally colored cottons are more commercially available. On the other hand, blue and yellow naturally colored cotton fibers are more difficult to find commercially. Color pigmentation occurs after the opening of the cotton ball. Color of the cotton fiber is white during the first week of the fiber ball opening, later as result of the sun shine genotypic constitution causes the color change of the fiber to the different shades of brown depending on the season of the year and the location due to climate and soil type [17–19]. It was stated that naturally brown colored cotton is very similar in morphology to common white cotton [20]. Although naturally colored cotton display the same structure as common white cotton, the pigmentation difference, existed at the center of the lumen, is the only difference between these two species [20–22]. A dominant gene (the genetic factor) and the environmental factors influence the color shade and its intensity of the naturally colored cotton fiber [16, 23, 24]. It was reported that the varied color shades of brown and red-brown are mostly because of catechin–tannins and protein–tannin polymers in the lumen [25]. It was also stated that the color shades of naturally colored cotton is most likely the product of a variation of pigments which are bound to cellulose by sugar related links together with lignins, tannins, and other non-cellulosic materials [22, 26]. Also, it was reported that flavonoid compounds, such as flavonone, flavonol and anthocyanidin may be the existing pigments in the naturally colored cotton fibers [5, 24].

The naturally colored cotton fibers can be used for the production of yarns leading to woven and knitted fabrics or for nonwoven production. Naturally colored cotton fibers are generally used in handicrafts, t-shirts, blankets, jackets, knitwears, sweaters, socks, towels, shirts, underwear, intimate apparels and other clothing articles, home decorations and furnishings [27]. However, it was stated that the handle of towels made from naturally colored cottons may be harsh, and their moisture absorbency might be lower than the expected levels which then may lead to customer complaints due to such comfort concerning issues [9]. This may be due to the substances such as fat, pectin and lignin etc. contained in the colored cotton leading to undesirable hydrophobic nature [9].

Naturally colored cotton, similar to common white cotton, is largely composed of cellulose along with some non-cellulosic components. Non-cellulosic components such as wax, pectin, and protein are primarily found in the cuticle layer and in the primary wall [22]. Indeed, cotton fiber contains roughly 10% of non-cellulosic substances by weight such as fat, waxes, pectic substances, organic acids, protein/nitrogenous substances, non-cellulosic polysaccharides, mineral matters and other unidentified compounds [28–32]. The presence of these non-cellulosic materials in the cotton fiber is a good thing for the cotton fiber itself while growing, since these impurities create a physical hydrophobic barrier that protects the cotton fiber from the different environmental conditions such as weather throughout growing [33, 34]. In addition to these natural inherent impurities, seed-coat fragments, aborted seeds, leaves, and twigs etc. could mechanically adhere to the cotton fiber during harvest and picking leading to further impurities called 'motes'. What is more, cotton fiber product could also be contaminated with machine oils, tars or grease during yarn production, knitting and weaving etc. [35]. Therefore, scouring process containing alkali is a purifying treatment of cellulosic textiles and generally applied to cotton fibers in order to remove these impurities and to reduce the amount of impurities sufficiently to increase the hydrophilicity and to obtain level and reproducible results in dyeing, printing and finishing operations [35]. The scouring process is generally carried out with NaOH at high temperature with or without pressure. In here, alkaline scouring process expedites saponification of the esters present in the cotton wax and neutralizes the free fatty acids in the cotton fiber and therefore; the used alkaline decreases the interfacial tension of the remaining wax.

Moreover, electrolytic dissociation of the cellulose might decrease the adhesion of oil substances to the cotton fiber surface [14, 36].

Scouring process is generally applied for common white cotton fibers. Although there may not be a need for coloration process for naturally colored cotton fibers, scouring is also needed for naturally colored cotton fibers due to their hydrophobic nature in order to increase the level of hydrophilicity. As earlier mentioned, customers expect high moisture absorption from many textile products such as towels, t-shirts and underwear. Therefore, some research studies have been carried out regarding the scouring and surface modification of naturally colored cotton fibers. Gu [9] studied the scouring process of naturally green colored cotton with NaOH solution in order to decrease the hydrophobic content and ascend its moisture absorbency. Green colored cotton exhibited higher moisture absorbency and decrease in the crystalline region after alkaline scouring process. Demir et al. [22] studied the effect of argon and air atmospheric plasma on naturally colored cotton knitted fabrics. They concluded that atmospheric-plasma treatments are capable of modifying the surface of naturally colored cotton fabrics without any significant loss in the color yield or color fastness and thermal characteristics. Kang and Epps [37] studied to improve low moisture regain of three naturally colored cottons with the effect of enzymatic process and scouring process. They concluded that the moisture regain of three naturally colored cottons rose following scouring process and ascended further after enzyme process. Also Tsaliki et al. [38] studied the effects of scouring treatment (with NaOH and NaCO$_3$) and enzyme process on the color fastness and CIE L^*, a^*, b^* properties of upland white and brown naturally pigmented cotton fibers. They concluded that wet treatment did not affect the color fastness of the studied brown cotton. What is more, enzymatic process rather developed the overall performance of the brown colored cotton. In another study, the effects of scouring, with sodium carbonate and sodium hydroxide, and enzyme treatment, with a mixture of pectinase and cellulase, on the color of three naturally colored cottons (buffalo brown, coyote brown, and green cotton) were investigated [39]. They concluded that an enzymatic scouring treatment did not considerably alter the color of studied naturally colored cottons. On the other hand, alkali scouring processes resulted in lower lightness and chroma leading to darker colors.

The common natural white and off-white cotton fiber and their fabrics still comprise naturally occurring coloring substances even after scouring treatment [35]. These coloring matters from white cotton fibers can be removed by oxidizing bleaching agents such as using hydrogen peroxide. Bleaching process with hydrogen peroxide is generally applied to white and off-white cotton fibers in order to increase their whiteness properties prior to further wet-processing such as coloration and/or finishing. This process is an inevitable step for a finished textile product. Bleached cotton fabrics can directly be used for apparel industry with their white appearance without any further coloration process. Or, one should apply a proper bleaching process in order to dye (color) the cotton fabrics properly and to reach the aimed color properties reliably. Not only white and off-white cotton fibers but also naturally colored cotton fibers may be exposed to bleaching operations according to their end-use applications. As earlier mentioned, naturally colored cotton could be used in many different applications. Indeed, naturally colored cotton fibers can be solely used in yarn or fabric structures or they can be blended not only with common white and off-white cotton fibers but also with other naturally colored cottons exhibiting different colors with lighter and darker shades. Moreover, naturally colored cottons can also be used in blends with other cellulose and protein fibers. This blending process could be carried out during fiber, yarn, fabric or even multilayer fabric production stages. Therefore, the net effects of common bleaching process utilizing hydrogen peroxide (H_2O_2) on the colorimetric and hydrophilicity performances of the 100% naturally colored cotton fibers should be known prior to their possible solely usage or their usage in the blends for different end-use applications.

There is one study regarding the blending of naturally colored cotton with white cotton with different amounts of blending. In here, Parmar et al. [7] blended camel brown and olive green naturally colored cottons with white cotton fibers during the fiber stage. Then, they had created twill woven cotton fabric samples from 11 s count 100% white cotton fiber as a warp yarn and 8 s count weft yarn which is composed from different amount of white cotton fiber and different amounts of naturally camel brown and olive green colored cotton fibers. So the final cotton fabrics have white cotton warp yarn and blended cotton fiber weft yarns with different blending amounts. The final fabrics possess 0, 16.7, 27.8, 39.1, and 55.67% naturally colored fibers in the woven fabric structure. And then those cotton fabrics were treated with various chemicals such as tannic acid, ferrous sulphate, potassium aluminum sulphate, copper sulphate and with their different combinations in order to create different colors and shades from naturally colored cotton and white cotton fibers combination. Also it was stated that hydrogen peroxide bleaching was applied to these colored and white cotton fibers blended cotton fabrics (possessing 0, 16.7, 27.8, 39.1, and 55.67% naturally colored fibers in the fabric structure) in order to investigate their whiteness property (in Hunter). They have only studied the whiteness properties of these fabrics. Unfortunately, there is also no information found in the paper about how the bleaching operation was carried out for these blended cotton fiber fabrics and/or the bleaching recipe was also not given in the paper [7]. It was concluded that all brown cotton fiber containing blended fabrics (possessing 0, 16.7, 27.8, 39.1, and 55.67% naturally brown colored fibers in the fabric structure) became almost white after bleaching process. And on the other hand, the whiteness property of fabrics comprising green cotton fibers (possessing 0, 16.7, 27.8, 39.1, and 55.67% naturally green colored fibers in the fabric structure) were in the range of 74–84 where whiteness property decreased with the rise in the amount of green colored cotton fiber in the fabric structure (from 84 Hunter with 16.7% green cotton fiber presence to 74 Hunter with 55.7% green cotton fiber presence) [7].

However, detailed studies regarding the effects of hydrogen peroxide bleaching on the colorimetric and hydrophilicity properties of solely 100% naturally colored cotton fibers (their alone usage without any other fiber involvement and without any blending) have not been found in the examined literature. Once more it is important to remind that organic cotton production does not consume most synthetically compounded chemicals (fertilizers, insecticides, herbicides, growth regulators and defoliants) which are suggested for only conventional cotton production. Consequently, organic cotton production results in much more environmentally cotton fiber production when compared to conventional cotton fiber growing. So, in this experimental study, not only colorimetric (CIE L^*, a^*, b^*, C^*, $h°$, K/S, and whiteness properties etc.) but also hydrophilicity properties of studied two white (Nazilli 84 S and Aydın 110) and three naturally colored (Emirel, Akdemir, Nazilli DT-15) unique Turkish organic cotton fiber types were investigated before and after scouring (with NaOH), conventional bleaching (with H_2O_2) and the combination application of scouring and bleaching (scouring + bleaching) processes. The road map of this study is to explore and compare the changes in the colorimetric and hydrophilicity properties of studied these white and naturally colored organic cotton fibers, cultivated in Turkey, after pre-treatment processes in detail when compared to greige (un-treated) organic cotton fibers.

2   Materials and Methods

2.1   Cotton Fiber Cultivation

Materials

Two white (Nazilli 84 S and Aydın 110) and three naturally colored (Emirel, Akdemir, Nazilli DT-15) unique Turkish cotton fiber types were selected and cultivated, in line with the organic cotton fiber production, for this study under the control of the Turkey Nazilli Cotton Research Institute in Turkey. Cumhuriyet 75 wheat variety was used as a rotation plant during the cotton fiber cultivation. These utilized white and naturally colored unique Turkish cotton fibers are all *Gossypium hirsutum* L. and were developed via crossbreeding and selective breeding techniques in Turkey. Visual appearances the used organic cotton fiber types are given in the Table 1. The colors of the studied three naturally colored organic cotton fibers are camel hair color (Nazilli DT-15) and brown color (Emirel and Akdemir) (Table 1). Two white (Nazilli 84 S and Aydın 110) and three naturally colored (Emirel, Akdemir, Nazilli DT-15) unique Turkish cotton fiber types were planted and cultivated (in accordance with the organic cotton fiber production rules under the control of the Turkey Nazilli Cotton Research Institute) in two different plantation locations in the Aegean region of Turkey [Menemen/İzmir (Bakırçay Basin) and Sarayköy/Denizli (Büyük Menderes Basin) locations]. Studied organic cotton fiber types were the blend (mixture) of the respective organic cotton fibers from two different plantation locations.

2.2   Pre-treatment Processes

Scouring process (with NaOH), conventional bleaching process (with H_2O_2) and the combination application of scouring and bleaching processes together (scouring +

Table 1 Visual appearances of used Turkish organic cotton fiber types

Registered name of the organic cotton fiber	Nazilli 84 S	Aydın 110	Nazilli DT 15	Emirel	Akdemir
Visual appearance					

Table 2 Application conditions of implemented scouring and bleaching processes to the studied organic cotton fibers

Application conditions	Scouring	Bleaching
Concentrations	1 ml/l sequestering agent 1 ml/l non-ionic wetting agent % 2 caustic soda (NaOH)	2.5 g/l caustic soda 2.5 ml/l hydrogen peroxide (50%) (H_2O_2) 2 g/l non-ionic wetting agent 1 g/l stabiliser
Temperature (°C)	90	90
Time (minutes)	30	60
Liquor ratio	1/25	1/25
After treatment (washings)	Rinsing at 50 °C for 2 min then cold washing for 2 min	Rinsing at 50 °C for 2 min then cold washing for 2 min then neutralization with 1 ml/l acetic acid for 2 min afterwards cold washing for 1 min

bleaching) were applied to studied two white (Nazilli 84 S and Aydın 110) and three naturally colored (Emirel, Akdemir, Nazilli DT-15) unique Turkish organic cotton fiber types in order to examine their colorimetric (CIE L^*, a^*, b^*, C^*, $h°$, K/S, and whiteness properties etc.) and hydrophilicity properties before and after various wet pre-treatments. The application recipes of implemented scouring and bleaching processes for studied organic cotton fibers are shown in Table 2. In this study, there are three different types of pre-treatment processes. First one is only scouring process with sodium hydroxide. The second one is a direct bleaching process with hydrogen peroxide without any scouring process involvement (directly from greige to bleached). The third process type is the combination sequential usage of scouring and bleaching processes (scouring then bleaching = scouring + bleaching).

Scouring and bleaching processes of cotton fibers were carried out using Ataç Lab Dye HT model IR sample dyeing machine via the exhaustion process. After the stated washing cycles, fibers were flat-air-dried. Then, the changes in the colorimetric properties, whiteness degrees and hydrophilicity properties of studied white and naturally colored organic cotton fibers were determined after these pre-treatment processes in detail in comparison with their greige cotton fiber counterparts.

2.3 Whiteness Determination

Following the stated pre-treatment processes, the whiteness degree (measured in Stensby) of the studied organic cotton fibers was determined using a DataColor 600 spectrophotometer (DataColor International, Lawrenceville, NJ, USA). Each sample was measured from four different areas, twice on each side of the fibers for consistency, and the average value was calculated and reported.

2.4 Colorimetric Measurements

The CIE L^*, a^*, b^*, C^*, and $h°$ co-ordinates were measured and the K/S (*Color strength*) values calculated from the reflectance values at the appropriate wavelength of maximum absorbance (λ_{max}) for organic cotton fiber samples using a DataColor SpectraFlash 600 (DataColor International, Lawrenceville, NJ, USA), spectrophotometer (D65 day light, 10° standard observer). Each fiber sample was read in four different areas, twice on each side of the fibers for consistency, and the average value was calculated and reported.

2.5 Hydrophilicity Determination

The effectiveness of applied different pre-treatment processes on the hydrophilicity character of the studied white and naturally colored organic cotton fibers was detected through a sinking (submersion) test according to the TSE 866 test standard. The sinking test hydrophilicity determination was carried out under standard laboratory conditions (20 ± 2 °C, 65 ± 2% Relative Humidity). And all organic cotton fiber samples were conditioned under these conditions for 24 h prior to sinking test. The weight of the fiber specimens was 0.1 g. The samples were allowed to free fall 1 cm above the water surface of the water container which possesses distilled water at 21 ± 2 °C. The time of the submersion (in seconds) into the water was reported as hydrophilicity level. Ten measurements were carried out for each organic cotton fiber sample and the average value was calculated and reported in seconds. The shorter the submersion (sinking) time, the better the hydrophilicity which then leads to more hydrophilic cotton fiber.

3   Results and Discussion

3.1   Color Properties of Cotton Fibers Before and After Pre-treatment Processes

The reflectance spectra of studied greige (un-treated) Nazilli 84 S and Aydın 110, Emirel, Akdemir, Nazilli DT-15 organic cotton fiber types are shown in Fig. 1. Light is absorbed and reflected to a different degree at different wavelengths according to the color shade of the cotton fiber. It is clearly visible from the Fig. 1 that Nazilli 84 S and Aydın 110 are white organic cotton fibers with the highest reflectance values at each wavelength. It can be easily seen from the reflectance curves that Nazilli DT-15, Akdemir and Emirel cotton fibers are colored cotton fibers. Akdemir organic cotton fiber type is the darkest cotton fiber studied with the lowest reflectance values at each wavelength (Fig. 1). Moreover, Emirel organic cotton fiber is darker than Nazilli DT-15 which then makes Nazilli DT-15 the lightest (with the highest reflectance values) colored organic cotton fiber amongst the studied naturally colored organic cotton fibers.

The effect of scouring process on the reflectance spectra of naturally colored organic cotton fibers can be clearly seen in Fig. 2. The reflectance curves of scoured naturally colored organic cotton fibers are at lower level than those of greige (un-treated) counterparts. It is understood that after the scouring treatment, scoured naturally colored organic cotton fibers reflect less light and therefore their color becomes

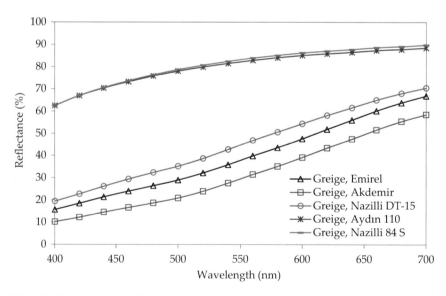

Fig. 1 Reflectance spectra of studied greige (un-treated) organic cotton fibers

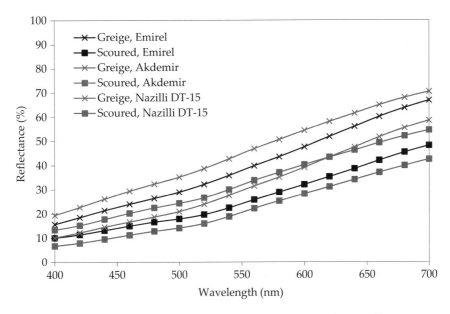

Fig. 2 Reflectance spectra of greige and scoured naturally colored organic cotton fibers

darker. Scoured colored organic cotton fibers exhibited consistently lower reflectance values than their greige counterparts at each wavelength (Fig. 2).

Whiteness Levels

Figure 3 displays the whiteness degrees, in Stensby, of studied white organic cotton fibers (Aydın 110 and Nazilli 84 S) before and after applied pre-treatment processes. After applied scouring process, whiteness degrees of both organic cotton fibers slightly increased. These slight whiteness increases can be expected. Since, as earlier mentioned, scouring process is generally applied to cotton fibers in order to remove and/or reduce impurities such as natural inherent impurities (non-cellulosic substances such as fat, waxes, pectic substances, organic acids, protein/nitrogenous substances, non-cellulosic polysaccharides, mineral matters and other unidentified compounds etc.), seed-coat fragments, aborted seeds, leaves, twigs etc. [14, 28–32]. The whiteness degree of cotton fibers slightly increased due to the removal of aforementioned foreign substances and impurities from the cotton fibers during the scouring process. It is known that alkaline scouring process expedites saponification of the esters present in the cotton wax and neutralizes the free fatty acids in the cotton fiber and therefore; the used alkaline decreases the interfacial tension of the remaining wax. In addition, electrolytic dissociation of the cellulose might decrease the adhesion of oil substances to the cotton fiber surface [14, 36]. Moreover, some small amount of the natural coloring matters in the white and off-white cotton fibers also might be deteriorated during this process due to alkaline action of the scouring at high temperature.

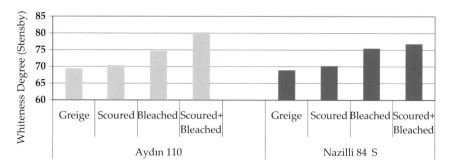

Fig. 3 Effect of pre-treatment processes on the whiteness properties of studied white organic cotton fibers

As expected, direct bleaching without any scouring process resulted in significantly higher whiteness levels than scouring process (Fig. 3). However, the highest whiteness increase was measured after the combination sequential usage of scouring and bleaching processes (scouring then bleaching = scouring + bleaching). It is known that the common natural white and off-white cotton fibers still comprise naturally occurring coloring substances even after scouring treatment [28]. These inherent coloring matters from white and off-white cotton fibers can be removed by hydrogen peroxide leading to significantly whiter appearance with higher whiteness degrees.

Color Properties

a^*–b^* and L^*–C^* plots and colorimetric data of studied greige naturally colored organic cotton fibers are given in Fig. 4 and Table 3, respectively. It is clearly visible from Table 3 and Fig. 4 that colors and shades of the studied naturally colored organic cotton fibers are slightly different from each other. Although the hue angle values measured ($h°$) of fibers were all below 90° (yellow-red axis zone) (Table 3), leading to brownish-reddish-yellowish shade appearance, naturally colored organic cotton fibers displayed slight different hue angles ($h°$) due to their different a^* and b^* values (Table 3; Fig. 4) leading to slight shade differences. These slight shade differences are clearly visible from the visual appearances of the fibers (Table 1).

The redness (a^*) and yellowness (b^*) values of greige (un-treated) naturally colored cotton fibers were slightly different (Fig. 4). Greige Akdemir organic cotton fiber is slightly redder and yellower in comparison to greige Emirel and greige Nazilli DT-15 organic cotton fibers due to its higher a^* and b^* values, respectively. On the other hand, greige Nazilli DT-15 exhibited the lowest a^* and b^* values leading to slightly less red and slightly less yellow appearance in comparison to other two fibers.

Moreover, greige Akdemir organic cotton fiber displayed the highest chroma value (C^*: 28.24) owing to its higher a^* and b^* values (Table 3; Fig. 4). Oppositely, greige Nazilli DT-15 exhibited the lowest chroma value (C^*: 23.60) due to its lower a^* and b^* values (Table 3; Fig. 4). Greige Akdemir organic cotton fiber displayed the lowest lightness (L^*: 61.82, Fig. 4b) value amongst greige naturally colored cotton

Fig. 4 **a** *a*–b** and **b** *L*–C** plots of studied greige (un-treated) naturally colored organic cotton fibers

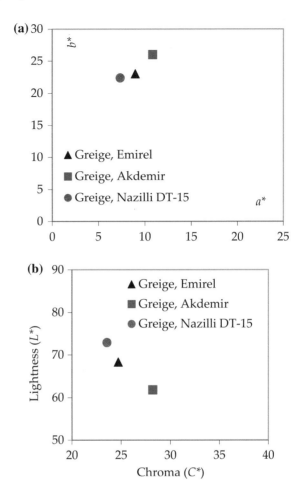

fibers leading to the highest color strength value (K/S: 3.98, Table 3), as expected. Conversely, greige Nazilli DT-15 exhibited the highest lightness (*L**: 72.92, Fig. 4b) value leading to the lowest color strength value (K/S: 1.67, Table 3). Colorimetric values of greige Emirel organic cotton fiber were in between those of other studied two naturally colored greige organic cotton fibers (Akdemir and Nazilli DT-15).

The color of greige Emirel organic cotton fiber is stronger than the color of greige Nazilli DT-15 due its lower lightness value (*L**) and higher chroma value (*C**) (Fig. 4b). But, the strongest color shade was measured for greige Akdemir organic cotton fiber owing to it's the lowest lightness and the highest chroma values leading to darker and more saturated appearance (Fig. 4b). From the other side, greige Nazilli DT-15 organic cotton fiber is the lightest and the least saturated color shade amongst greige naturally colored studied organic cotton fibers due to it's the highest *L** value and the lowest *C** value (Fig. 4b).

Table 3 Color properties of naturally colored organic cotton fibers before and after pre-treatment processes

Colored cotton type	Pre-treatment type	L^*	a^*	b^*	C^*	$h°$	K/S
Emirel	Greige (un-treated)	68.35	9.04	23.05	24.76	68.59	2.27
	Scoured	56.96	10.39	21.40	23.79	64.1	4.03
	Bleached	69.55	9.25	23.70	25.44	68.68	2.12
	Scoured + bleached	70.23	8.65	23.70	25.23	69.96	2.06
Nazilli DT-15	Greige (un-treated)	72.92	7.41	22.41	23.60	71.71	1.67
	Scoured	63.61	8.49	21.92	23.51	68.84	2.81
	Bleached	71.97	8.02	22.54	23.92	70.42	1.73
	Scoured + bleached	76.08	6.33	22.11	23.00	74.04	1.34
Akdemir	Greige (un-treated)	61.82	10.90	26.05	28.24	67.29	3.98
	Scoured	53.1	11.07	24.68	27.05	65.84	6.47
	Bleached	55.54	11.25	22.29	24.96	63.21	4.51
	Scoured + bleached	64.11	10.73	25.83	27.97	67.44	3.30

So overall, greige (un-treated) Akdemir naturally colored organic cotton fiber displayed the reddest (with the highest a^* value), the yellowest (with the highest b^* value) appearance, the highest chroma (the most saturated), the lowest lightness (the darkest) and the highest color strength (the strongest color yield) and therefore the strongest color shade amongst the studied greige colored cotton fibers (Fig. 4; Table 3). Oppositely, the greige (un-treated) Nazilli DT-15 organic cotton fiber exhibited the lowest redness (with the lowest a^* value), the lowest yellowness (with the lowest b^* value) appearance, the lowest chroma (the least saturated), the highest lightness (the lightest) and the lowest color strength (the weakest color yield) and therefore the weakest color shade amongst the studied greige colored organic cotton fibers (Fig. 8; Table 3). The colorimetric performance of greige (un-treated) Emirel naturally colored organic cotton fiber was in between the colorimetric properties of greige (un-treated) Akdemir and Nazilli DT-15 naturally colored organic cotton fibers.

The colorimetric properties of studied naturally colored organic cotton fibers before and after pre-treatment processes are given in Table 3 and Figs. 5, 6, 7, 8.

The lightness (L^*) values of all three naturally colored organic cotton fibers congruously decreased after scouring process leading to darker appearance (Fig. 5). As expected, in parallel, the color strength (K/S) values of all three naturally colored organic cotton fibers concertedly increased after scouring process leading to again darker appearance (Fig. 6). So, scouring process led to darker appearance with higher color strength on all studied naturally brownish colored organic cotton fibers (2.27 versus 4.03 for Emirel; 1.67 versus 2.81 for Nazilli DT-15; 3.98 versus 6.47 for Akdemir, respectively). It is right point to mention that scouring bath solutions after the scouring process were also strongly brownish colored. These results were actually in line with the study of Kang et al. [39] where the color change mech-

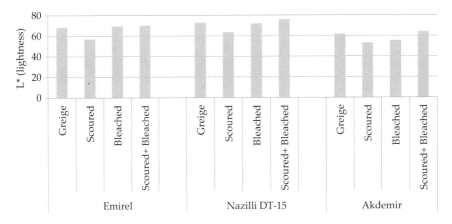

Fig. 5 Effect of pre-treatment processes on the lightness (L^*) properties of studied naturally colored organic cotton fibers

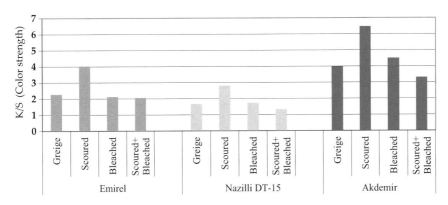

Fig. 6 Effect of pre-treatment processes on the color strength (K/S) properties of studied naturally colored organic cotton fibers

anisms of colored cotton fibers (buffalo brown, coyote brown, and green cotton) after alkali scouring (sodium carbonate and sodium hydroxide) and enzymatic treatment (mixture of pectinase and cellulase) were examined via wax extraction, light microscopy, scanning electron microscopy (SEM), X-ray analysis, and inductively coupled plasma (ICP) analysis. This study reported that the color of naturally colored cotton fibers exhibited deeper and darker appearance, and moreover the scouring solutions were also deeply colored. It was stated that after alkali scouring process, naturally colored cotton fibers became swollen and some of the inherent pigments in the naturally colored cottons moved towards the outer layer of the cotton fiber leading to darker appearance [39]. And moreover, some of the migrated pigments moved to the outside of the cotton fiber and departed from the cotton fiber into the scouring

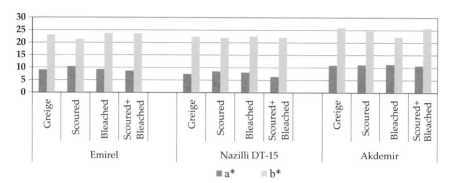

Fig. 7 Effect of pre-treatment processes on the a^* (redness/greenness)–b^* (yellowness/blueness) values of studied naturally colored organic cotton fibers

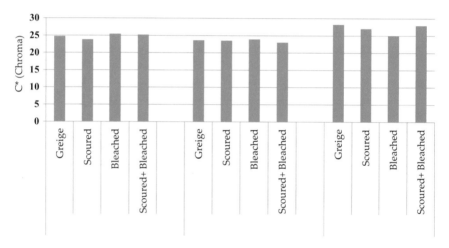

Fig. 8 Effect of pre-treatment processes on the C^* (chroma) values of studied naturally colored organic cotton fibers

bath making it strongly brown colored. During this process, also some potassium loss realized from the naturally colored cotton fibers [39].

In comparison to scoured organic cotton fibers, bleached and scoured + bleached organic cotton fibers exhibited higher lightness (L^*) and lower color strength (K/S) values leading to lighter appearance (Figs. 5 and 6). It is known that bleaching agents such as utilized hydrogen peroxide can oxidize or reduce coloring matters and inherent pigments in the fiber [35]. So the color intensity is decreased in here following the bleaching process when compared to scouring process alone. Although the color strength and lightness values of bleached and scoured + bleached organic cotton fibers were close and comparable to those of greige (un-treated) counterparts, the combination sequential usage of scouring and then bleaching processes (scouring + bleaching) resulted in very slightly lower color strength and slightly higher lightness values in comparison to greige un-treated counterparts leading to very slightly lighter appearance (Figs. 5 and 6). For instance, the color strengths of greige (un-treated) and scoured + bleached Emirel organic cotton fiber were 2.27 and 2.06, respectively. Similarly, the color strengths of greige (un-treated) and scoured + bleached Nazilli DT-15 organic cotton fiber were 1.67 and 1.34, respectively. Finally, the color strengths of greige (un-treated) and scoured + bleached Akdemir organic cotton fiber were 3.98 and 3.30, respectively. Nevertheless, greige (un-treated), bleached and scoured + bleached naturally colored organic cotton fibers displayed close color strength levels (2.27, 2.12, 2.06 for Emirel; 1.67, 1.73, 1.34 for Nazilli DT-15; 3.98, 4.51, 3.30 for Akdemir, respectively).

As expected from the changes on the lightness and color strength values, in all three naturally colored organic cotton fibers, a^* values were slightly increased and b^* values were slightly decreased after scouring process leading to slightly redder and slightly less yellow appearances. Similar observation regarding a^* level change was also experienced in a study of Tsaliki et al. [38] where stated that scouring process resulted in slight a^* value increase on brown cotton. a^* values of un-treated (greige), bleached and scoured + bleached Emirel naturally colored organic cotton fibers were close to each other and 9.04, 9.25 and 8.65 (Fig. 7; Table 3). Similarly, a^* values of un-treated (greige), bleached and scoured + bleached Nazilli DT-15 naturally colored organic cotton fibers were close to each other and 7.41, 8.02 and 6.33. In the case of Akdemir naturally colored organic cotton fiber, in parallel, a^* values of un-treated (greige), bleached and scoured + bleached fibers were close to each other and 10.90, 11.25 and 10.73 (Fig. 7; Table 3). b^* values of un-treated (greige), bleached and scoured + bleached Emirel naturally colored organic cotton fibers were very close to each other and 23.05, 23.70 and 23.70 (Fig. 7; Table 3). Similarly, b^* values of un-treated (greige), bleached and scoured + bleached Nazilli DT-15 naturally colored organic cotton fibers were very close to each other and 22.41, 22.54 and 22.11. In the case of Akdemir naturally colored organic cotton fiber, in parallel, b^* values of un-treated (greige), bleached and scoured + bleached fibers were close to each other and 26.05, 22.29 and 25.83 (Fig. 7; Table 3). Scoured + bleached colored organic cotton fibers were slightly less red in comparison to their greige (un-treated) counterparts.

After scouring process, all three naturally colored organic cotton fibers were very slightly less saturated due to their slightly less chroma (C^*) values in comparison to greige (un-treated) counterparts (Fig. 8). On the other hand, there is no clear trend after bleaching and scouring + bleaching processes regarding the chroma values of studied fibers. Overall, all greige (un-treated), bleached and scoured + bleached naturally colored organic cotton fibers exhibited similar close comparable chroma values (Fig. 8).

Overall, solely bleaching process (without any prior scouring process) and combination sequential usage of scouring and bleaching processes (scouring then bleaching = scouring + bleaching) generally did not significantly affect the color properties of studied naturally colored organic cotton fibers leading to similar close colorimetric performance with their greige (un-treated) counterparts.

3.2 Hydrophilicity Properties of Cotton Fibers Before and After Pre-treatment Processes

The hydrophilicity (sinking time, in seconds) properties of studied white and naturally colored organic cotton fibers before and after pre-treatment processes are given in Table 4. All white and naturally colored organic cotton fibers studied were buoyant and did not sink in the water and therefore not hydrophilic (hydrophobic) due to their natural inherent impurities (non-cellulosic components) leading to hydrophobic nature (Table 4). As earlier mentioned, naturally colored cotton, similar to common white cotton, is largely composed of cellulose along with some non-cellulosic components. It is known that non-cellulosic components of the cotton fibers, roughly 10%, such as waxes, fats, pectic substances, organic acids, non-cellulosic polysaccharides, protein/nitrogenous substances, mineral matters and other unidentified compounds are primarily found in the cuticle layer and in the primary wall leading to the creation of physical hydrophobic barrier [22, 28–35]. After scouring process, all studied white and naturally colored cotton fibers sank into the water with all less than 9 s and therefore they became hydrophilic. Fats found in the cotton fibers treated with the scouring solution are saponifying [40]. The resulting soaps enable non-saponified parts to become in emulsion state. Moreover, pectin, protein, and other substances are broken down into smaller molecules, and these small molecules were passed into the scouring bath as sodium salts [40]. In another way of saying, alkaline scouring process expedites saponification of the esters present in the cotton wax and neutralizes the free fatty acids in the cotton fiber and therefore; the used alkaline decreases the interfacial tension of the remaining wax. Moreover, electrolytic dissociation of the cellulose might decrease the adhesion of oil substances to the cotton fiber surface [14, 36]. In this way, cotton fibers gain hydrophilic character after scouring process. Also Gu [9] stated that molecules of fat and lignin move faster with greater energy during the scouring process leading to their liberation from the fiber body therefore increase on its moisture regain (%). In one way or another, it is clear that during the scouring process, some foreign substances and impurities of the studied cotton fibers are broken, removed and/or reduced leading to the formation of hydrophilic character.

Overall, the hydrophilicity values of the hydrophobic greige (un-treated) organic cotton fibers decreased to 6–8 s after scouring, 4–5 s after bleaching, and 3–4 s after scouring + bleaching combination leading to improvement in hydrophilicity levels of studied cotton fibers (Table 4). The highest hydrophilicity performance improvement was measured on organic cotton fibers treated with the combination sequential usage of scouring and bleaching processes (scouring then bleaching = scouring + bleaching). It seems that during the bleaching process, little bit more foreign substances and impurities of the studied organic cotton fibers are removed and/or reduced leading to slightly more hydrophilic character formation.

Table 4 Hydrophilicity (sinking time) properties of naturally colored cotton fibers before and after pre-treatment processes

Fiber type	Pre-treatment process type	Hydrophilicity-sinking time (seconds)
Aydın 110	*Greige (un-treated)*	Buoyant
	Scoured	7.12
	Bleached	5.08
	Scoured + bleached	3.32
Nazilli 84 S	*Greige (un-treated)*	Buoyant
	Scoured	6.71
	Bleached	4.26
	Scoured + bleached	3.09
Emirel	*Greige (un-treated)*	Buoyant
	Scoured	8.07
	Bleached	4.61
	Scoured + bleached	3.64
Nazilli DT-15	*Greige (un-treated)*	Buoyant
	Scoured	8.29
	Bleached	5.05
	Scoured + bleached	4.05
Akdemir	*Greige (un-treated)*	Buoyant
	Scoured	8.43
	Bleached	5.25
	Scoured + bleached	3.59

4 Conclusions

The cultivation of naturally colored organic cotton has captured the attention lately due to the increasing environmental concerns and positive attributes of naturally colored cotton fibers. Naturally colored cotton fibers are generally utilized in hand-icrafts, t-shirt, blanket, jacket, knitwear, sweaters, socks, towels, shirts, underwear, intimate apparels and other clothing articles, home decorations and furnishings etc. Non-cellulosic components of the cotton fibers, roughly 10%, such as waxes, fats, pectic substances, organic acids, non-cellulosic polysaccharides, protein/nitrogenous substances, mineral matters and other unidentified compounds are primarily found in the cuticle layer and in the primary wall leading to the creation of physical hydrophobic barrier. Although there may not be a need for dyeing process for naturally colored cotton fibers, scouring is also needed for naturally colored cotton fibers due to their hydrophobic nature in order to increase the level of hydrophilicity. Since, customers generally expect high moisture absorption from many textile products such as towels, t-shirts and underwear. Moreover, not only white and off-white cotton fibers but also naturally colored cotton fibers may be exposed to bleaching operations according to their end-use applications. Since, naturally colored cotton fibers can be solely used in yarn or fabric structures or they can be blended (during fiber, yarn, fabric or even multilayer fabric production stages) not only with common white and off-white cotton fibers but also with other naturally colored cottons exhibiting different colors with lighter and darker shades. What is more, naturally colored cottons can also be used in blends with other cellulose and protein fibers. Therefore, the net effects of common bleaching process utilizing hydrogen peroxide (H_2O_2) on the colorimetric and hydrophilicity performances of the 100% naturally colored cotton fibers should be known prior to their possible solely usage or their usage in the blends for different end-use applications. Therefore, colorimetric (CIE L^*, a^*, b^*, C^*, $h°$, K/S, and whiteness properties etc.) and hydrophilicity properties of studied two white (Nazilli 84 S and Aydın 110) and three naturally colored (Emirel, Akdemir, Nazilli DT-15) unique Turkish organic cotton fiber types will also be investigated before and after scouring (with NaOH), conventional bleaching (with H_2O_2) and the combination application of scouring and bleaching (scouring + bleaching) processes in comparison with their greige (un-treated) organic cotton fiber counterparts.

Greige (un-treated) Akdemir naturally colored organic cotton fiber displayed the reddest (with the highest a^* value), the yellowest (with the highest b^* value) appearance, the highest chroma (the most saturated), the lowest lightness (the darkest) and the highest color strength (the strongest color yield) and therefore the strongest color shade amongst the studied greige (un-treated) naturally colored organic cotton fibers. In contrast, the greige (un-treated) Nazilli DT-15 naturally colored organic cotton fiber exhibited the lowest redness (with the lowest a^* value), the lowest yellowness (with the lowest b^* value) appearance, the lowest chroma (the least saturated), the highest lightness (the lightest) and the lowest color strength (the weakest color yield) and therefore the weakest color shade amongst the studied greige colored organic cotton fibers. The colorimetric performance of greige (un-treated) Emirel naturally colored organic cotton fiber was in between the colorimetric properties of greige (un-treated) Akdemir and greige (un-treated) Nazilli DT-15 naturally colored organic cotton fibers.

After applied scouring process, whiteness degrees of both white organic cotton fibers slightly increased. As expected, the highest whiteness increase on white organic cotton fibers was measured after the combination sequential usage of scouring and bleaching processes (scouring then bleaching = scouring + bleaching). Since, inherent coloring matters from white and off-white cotton fibers can be broken up and/or removed by hydrogen peroxide leading to significantly whiter appearance with higher whiteness degrees.

Scoured naturally colored organic cotton fibers exhibited consistently lower reflectance values than their greige counterparts at each wavelength. What is more, after scouring process, all three naturally colored organic cotton fibers congruously exhibited darker [with the lower lightness (L^*) values and higher color strength (K/S) values], slightly redder (slightly higher a^* values) and slightly less yellow (slightly lower b^* values) appearance in comparison to their greige (un-treated) counterparts. This situation can be explained as follows that naturally colored cotton fibers became swollen and some of the inherent pigments in the naturally colored cottons moved towards the outer layer of the cotton fiber leading to darker appearance. In comparison to scoured organic cotton fibers, bleached and scoured + bleached organic cotton fibers exhibited higher lightness ($L*$) and lower color strength (K/S) values leading to lighter appearance. It is known that bleaching agents such as utilized hydrogen peroxide can oxidize or reduce coloring matters and inherent pigments in the fiber. On the other hand, greige (un-treated), bleached and scoured + bleached naturally colored organic cotton fibers displayed close color strength levels. Overall, it can be concluded that solely bleaching process (without any prior scouring process) and combination sequential usage of scouring and bleaching processes (scouring then bleaching = scouring + bleaching) generally did not significantly affect the color properties of studied naturally colored cotton fibers leading to similar close colorimetric performance with their greige (un-treated) counterparts. So, after the bleaching process, scoured naturally colored organic cotton fibers which darkened due to the scouring process roughly turned back to their original colorimetric levels of greige (un-treated) versions. In this case, if the naturally colored organic cotton fibers are blended with the normal white and off-white organic cotton fibers or other cellulosic fibers, applied bleaching process does not cause a significant color change in the naturally colored organic cotton fibers and this indicates that they will approximately remain at the same color property levels as their greige (un-treated) counterparts.

All white and naturally colored organic cotton fibers studied were buoyant and did not sink in the water and therefore not hydrophilic (hydrophobic) due to their natural inherent impurities (non-cellulosic components) leading to hydrophobic nature. After scouring process, all studied white and naturally colored organic cotton fibers sank into the water with all less than 9 s and therefore they became hydrophilic. It is clear that during the scouring process, some foreign substances and impurities of the studied cotton fibers are broken, removed and/or reduced leading to the formation of hydrophilic character. Overall, the hydrophilicity values of the hydrophobic greige (un-treated) organic cotton fibers decreased to 6–8 s after scouring, 4–5 s after bleaching, and 3–4 s after scouring + bleaching combination leading to improvement in hydrophilicity levels of studied organic cotton fibers. The highest hydrophilicity performance improvement was measured on organic cotton fibers treated with the combination sequential usage of scouring and bleaching processes (scouring then bleaching = scouring + bleaching). It seems that during the bleaching process, little bit more foreign substances and impurities of the studied organic cotton fibers are removed and/or reduced leading to slightly more hydrophilic character formation. So, the bleaching process following the scouring process slightly increases the hydrophilicity values of both white and naturally colored organic cotton fibers leading to more hydrophilic fibers. This measured hydrophilicity improvement is really important not only for the sake of further wet processing such as coloration and finishing processes etc. but also for the comfort and moisture management properties of textile apparel products made from organic cotton fibers.

References

1. Vreeland J, James M (1999) The revival of colored cotton. Scientific America, 280 (4), 112–119; Lee J (1996) A new spin on naturally colored cotton. Agri Res 44(4):20–21
2. Mohammadioun M, Gallaway M, Apodaca JK (1994) An economic analysis of organic cotton as a niche crop in texas. University of Texas at Austin, Bureau of Business Research, Austin, Texas
3. Burnett P (1995) Cotton naturally. Text Horiz 15:36–38
4. Murthy MSS (2001) Never say dye: the story of coloured cotton. Resonance 6:29–35
5. Hua S, Yuan S, Shamsi IH, Zhao X, Zhang X, Liu Y, Wen G, Wang X, Zhang H (2009) A comparison of three isolines of cotton differing in fiber color for yield, quality, and photosynthesis. Crop Sci 49(3):983–989
6. Reddy N, Yang Y (2015) Innovative biofibers from renewable resources. Springer, New York
7. Parmar MS, Sharma RP (2002) Development of various colours and shades in naturally coloured cotton fabrics. Indian J Fibre Text Res 27:397–407
8. Kimmel LB, Day MP (2001) New life for an old fiber: attributes and advantages of naturally colored cotton. Aatcc Rev 1(10)
9. Gu H (2005) Research on the improvement of the moisture absorbency of naturally self-coloured cotton, 2005. J Text Inst JOTI 96(4):247–250
10. Sundaramurthy VT (1994) Indian Text J 126–128; Sundaramurthy VT (1949) Heritable relationships of brown lints in cotton. Agronomy J 188–191
11. Vreeland JM Jr (1996) Organic and naturally coloured native cotton from Peru, New Research in Organic Agriculture, 11th International Scientific IFOAM Conference, August 11–15, Copenhagen
12. Parmar MS, Chakraborty M (2001) Thermal and burning behavior of naturally colored cotton. Text Res J 71(12):1099–1102
13. Hustvedt G, Crews C (2005) The Ultraviolet protection factor of naturally-pigmented cotton. J Cotton Sci 49:47–55
14. Kang SY (2007) Investigation of color change and moisture regain of naturally colored cotton. Doct Philos Athens, Georgia
15. Dutt Y, Wang XD, Zhu YG, Li1 YY (2004) Breeding for high yield and fibre quality in coloured cotton. Plant Breed 123:145–151 (2004). ISSN 0179-9541
16. de Morais Teixeira E, Correâ AC, Manzoli A, de Lima Leite F, de Oliveira CR, Mattoso LHC (2010) Cellulose nanofibers from white and naturally colored cotton fibers. Cellulose 17:595–606
17. Carvalho L, Farias F, Lima ve M, Rodrigues J (2014) Inheritance of different fiber colors in cotton (*Gossypium barbadense* L.). Crop Breed Appl Biotechnol 14(4):256–260
18. Chaudry M (1992) Natural colors of cotton. ICAC Recorder, Washington
19. Dickerson D, Lane ve D. Rodrigues E (1999) Naturally coloured cotton: resistance to changes in color and durability when refurbished with selected laundry aid. California State University, Agricultural Technology Institute
20. Elesini US, Cuden AP, Richards AF (2002) Study of the green cotton fibres. Acta Chim Slov 49:815–833
21. Vreeland JM (1999) In organic cotton, from field to final product. In: Myers D, Stolton S (eds) Intermediate Technology, London
22. Demir A, Özdoğan A, Özdil N, Gürel A (2011) Ecological materials and methods in the textile industry: atmospheric-plasma treatments of naturally colored cotton. J Appl Polym Sci 119:1410–1416
23. Carvalho LP, Dos Santos JW (2003) Respostas correlacionadas do algodoeiro com a selec¸ã̃o para a colorac¸ã̃o da fibra. Pesq Agropec Bras 38:79–83
24. Xiao YH, Zhang Z-S, Yin M-H, Luo M, Li X-B, Hou L, Pei Y (2007) Cotton flavonoid structural genes related to the pigmentation in brown fibers. Biochem Biophys Res Commun 358:73–78

25. Ryser U (1999) Cotton fiber initiation and histodifferentiation, in cotton fibers, developmental biology, quality improvement, and textile processing. Basra AS (ed) Haworth Press, Binghamton, New York, Chap. 1, pp. 21–29
26. Buschle G, Knight D, Knight C, Fox SV (1998) Proceedings beltwide cotton conferences, January 5–9, San Diego, California. National Cotton Council, 1:731
27. Vreeland JM (1999) Organic cotton: from field to final product (Myers D and Stolton S as editors). Intermediate Technology Publications, London, Guildford
28. Karmakar SR (1999) Application in the pre-treatment processes of textiles. Colourage Annu 75–84:121
29. Li YH, Hardin IR (1998) Enzymatic scouring of cotton—surfactants, agitation, and selection of enzymes. Text Chem Colorist 30(9):23–29
30. Li YH, Hardin IR (1998) Treating cotton with cellulases and pectinases: effects on cuticle and fiber properties. Text Res J 68(9):671–679
31. Li YH, Hardin IR (1997) Enzymatic scouring of cotton: effects on structural properties. Text Chem Colorist 29(8):71–76
32. Lin CH, Hsieh YL (2001) Direct scouring of greige cotton fabrics with proteases. Text Res J 71(5):425–434
33. Hartzell-Lawson M, Durrant SK (2000) The efficiency of pectinase scouring with agitation to improve cotton fabric wettability. Textile Chem Color Am Dyest Repor 32(8)
34. Etters JN (1999) Cotton preparation with alkaline pectinase: an environmental advance. Text Chem Colorist Am Dyestuff Reporter 1(3):33–36
35. Karmakar SR (1999) Chemical technology in the pre-treatment processes of textiles. vol 12, Elsevier
36. Valko EI (1955) Bleaching. In: Ward K Jr (ed) Chemistry and chemical technology of cotton. Interscience Publishers, New York, pp 121, 152–156
37. Kang SY, Epps HH (2009) Effect of scouring and enzyme treatment on moisture regain percentage of naturally colored cottons. J Text Inst 100(7):598–606
38. Tsaliki E, Kechagia U, Eleftheriadis I, Xanthopoulos F (2016) Effect of wet treatments on color fastness of naturally colored cotton. J Nat Fibers 13(4):502–505
39. Kang SY, Epps HH (2008) Effect of scouring on the color of naturally-colored cotton and the mechanism of color change. AATCC Rev 8(7):38–43, 6p
40. Tarakçıoğlu I (1979) Tekstil Terbiyesi ve Makinaları Cilt 1, Tekstil Terbiyesinde Temel İşlemker ve Selüloz Liflerinin Terbiyesi. Ege Üniversitesi Tekstil Fakültesi Yayınları No: 2, Ege Üniversitesi Matbaası, İzmir, Bornova

Physical Properties of Different Turkish Organic Cotton Fiber Types Depending on the Cultivation Area

Sema Palamutcu, Ali Serkan Soydan, Ozan Avinc,
Gizem Karakan Günaydin, Arzu Yavas, M. Niyazi Kıvılcım
and Mehmet Demirtaş

Abstract Cotton fiber properties are one major issue to estimate the sale price, to optimize the production process of the highest yarn quality with the lowest level of fiber lost. Fiber properties are influenced by fiber genetic codes, growing conditions of humidity, temperature, and soil content of the land. In this study, examined white and naturally colored unique Turkish cotton fibers were developed via crossbreeding and selective breeding techniques in Turkey. The physical properties [fineness (micronaire index), fiber length (mm), and fiber strength (g/tex)] of different varieties of two white (Nazilli 84 S and Aydın 110) and three naturally colored (Emirel, Akdemir, Nazilli DT-15) organic cotton fiber types which cultivated (in compliance with the organic cotton fiber production under the control of the Turkey Nazilli Cotton Research Institute) in two different plantation locations in the Aegean Region of Turkey [Menemen/İzmir (Bakırçay Basin) and Sarayköy/Denizli (Büyük Menderes Basin) locations] under different climate types and weather conditions for five consecutive year period (from 2012 to 2016) were investigated. Measured and recorded data are analyzed with using a statistical evaluation method of Least Squares Fit model to accomplish Analysis of Variance and Effect Tests. Statistical evaluation has been designed to evaluate influence of dependent variable of fiber type, location, and year on the independent fiber properties of length, strength and fineness (micronaire). Different type of organic cotton fibers, different weather conditions of crop year and different cultivation location are found somehow influential factors on the studied major fiber properties. Statistical evaluation of the fiber length has shown that length changes depending on the crop year and location differences where seasonal weather conditions vary. In the case of fiber fineness, especially crop year is found the most influential factor that the same fiber exhibited different fiber fineness

S. Palamutcu · A. S. Soydan · O. Avinc · A. Yavas (✉)
Textiles Engineering Department, Pamukkale University, Denizli 20016, Turkey
e-mail: aozerdem@pau.edu.tr

G. K. Günaydin
Buldan Vocational School, Pamukkale University, Buldan, Denizli, Turkey

M. Niyazi Kıvılcım · M. Demirtaş
Cotton Research Institute, Nazilli, Aydın, Turkey

© Springer Nature Singapore Pte Ltd. 2019
M. A. Gardetti and S. S. Muthu (eds.), *Organic Cotton*, Textile Science
and Clothing Technology, https://doi.org/10.1007/978-981-10-8782-0_2

values depending on the slight weather condition changes of year and also location. Seasonal climate differences of different years are found the most influential factor on the fiber strength. Organic cotton fiber type of Aydın 110, which is white organic cotton fiber, is found having the best fiber properties among the five fiber types, and naturally colored organic cotton fiber type of Akdemir, which is one of the studied naturally colored organic cotton fibers, is found having the lowest fiber properties in general.

Keywords Organic cotton · Cotton cultivation · Turkish cotton · Colored cotton Strength · Fineness · Length

1   Introduction

Cotton fiber cultivation lines are major economic activities in the frame of agricultural and industrial activities in Turkey. Turkish Republic has been established in the lands of historically known Asia Minor where cotton fiber and its processing as textile and clothing items are historically and traditionally very well practiced for the long time. Cotton fiber is currently one of the major agriculture plants in modern Turkey where about 500,000 ha lands in average have been used for cotton cultivation and about 850,000 tons of crop in average is harvested [1]. Turkey is one of the major cotton fiber grower countries in the world with the 8th place after other main cotton fiber grower countries of India, China, USA, Pakistan, Brazil, Uzbekistan, and Australia [2]. Organic cotton fiber can be defined as "more sustainable" than the conventional cotton fiber which is an environmentally preferable product [3]. Since, organic cotton fiber production does not consume most synthetically compounded chemicals (fertilizers, insecticides, herbicides, growth regulators and defoliants) which are advised for only conventional cotton production [3–5]. Turkey has well established organic cotton regions and organic cotton fiber yields are considerably high in Turkey.

Leading researches about cotton fiber types are continued in Turkey and in the world in order to increase the cotton yield per acre and moreover to improve fiber properties and to adapt fiber types to the different type of soil and weather conditions. Different weather conditions on the cultivation location could affect the final fiber properties. Moreover, cotton fiber breeders are also working to improve the properties and quality of different types of cotton fiber genotypes (white, off-white and naturally colored) by using the conventional plant breeding techniques such as a selection, crossing and mutation and utilizing biotechnology leading to the production of better and more productive superior varieties of cotton fibers. These researches can be conducted in the governmental institutions, universities, and private research companies in the world. In Turkey, cotton fiber studies are managed in governmental and semi-governmental institutes and universities. The Turkish Ministry of Food, Agriculture and Livestock-(MFAL) is the main state organization involved in agriculture and rural development in Turkey. The Ministry coordinates and implements the agricultural R&D activities where cotton fiber research works are conducted by the

Cotton Research Institute (Cotton Research Station), in Nazilli-Aydin-Turkey, being the primary mono crop multidisciplinary research institute. Nazilli Cotton Research Institute is a well-established one of the older foundation that works as cotton fiber improving branch of the Republic of Turkey Ministry of Food Agriculture and Livestock which has been investigating on conserving of genetic stock, developing source material, breeding, collection and evaluation of data on cotton in National Scale since 1934. Most of these aforementioned researches have been carried out throughout the world for the common white and off-white cotton fibers. Although common white cotton fiber is the most widely known and used type of cotton fiber for years all around the world, there is also naturally colored types of cotton fiber in the world.

Once more, it is vital to remind that cotton fiber properties are one major issue to estimate the sale price, to optimize the production process of the highest yarn quality with the lowest level of fiber lost. Fiber properties are influenced by fiber genetic codes, growing conditions of humidity, temperature, and soil content of the land. In this experimental study, the utilized white and naturally colored unique Turkish organic cotton fibers were developed via crossbreeding and selective breeding techniques in Turkey. The physical properties (fiber length, fiber fineness, fiber strength etc.) of different varieties of Turkish organic cotton fiber types (including white, naturally colored organic cotton and hybrid cottons) which cultivated in two different parts of Turkey (under different climate types) for five-year period were investigated and discussed. Five unique types of white and naturally colored organic cotton fibers that are grown, in line with the organic cotton fiber production, under the control of the Nazilli Cotton Research Institute of Turkey in two different locations are used to evaluate their physical and technological fiber properties according to actual weather conditions of the cultivated locations from 2012 to 2016. Since, the fiber properties of length, strength and fineness are most influential cotton fiber properties on the yarn processing efficiency and yarn properties. Fiber cultivation for all these five different cotton fibers has been carried out in two different locations (Menemen/İzmir and Sarayköy/Denizli locations) in the Aegean region of Turkey for aforementioned 5 years period. Weather conditions of both plantation locations have displayed different temperature and different relative humidity conditions as well as different amount of rain over these 5 years period. The road map of this study is to explore and compare the changes on physical properties [fineness (micronaire index), fiber length (mm), and fiber strength (g/tex)] of studied two white organic cotton fibers (Nazilli 84 S and Aydın 110) and three naturally colored organic cotton fibers (Emirel, Akdemir, Nazilli DT-15) with the statistical evaluation using Least Squares Fit model to accomplish Analysis of Variance and Effect Tests.

2 Materials and Methods

2.1 Cotton Fiber Cultivation

2.1.1 Materials

Two white (Nazilli 84 S and Aydın 110) and three naturally colored (Emirel, Akdemir, Nazilli DT-15) unique Turkish organic cotton fiber types were selected and cultivated for this study under the control of the Turkey Nazilli Cotton Research Institute in Turkey in line with the organic cotton fiber production. Cumhuriyet 75 wheat variety was used as a rotation plant during the cotton fiber cultivation. Detailed information about the used organic cotton fiber types and their visual appearances are given in Table 1. The colors of the studied three naturally colored organic cotton fibers are camel hair color and brown color (Table 1).

2.1.2 Cultivation Locations and Their Weather Conditions

Two white (Nazilli 84 S and Aydın 110) and three naturally colored (Emirel, Akdemir, Nazilli DT-15) unique Turkish organic cotton fiber types were planted and cultivated (in compliance with the organic cotton fiber production under the control of the Turkey Nazilli Cotton Research Institute) in two different plantation locations in the Aegean region of Turkey [Menemen/İzmir (Bakırçay Basin) and Sarayköy/Denizli (Büyük Menderes Basin) locations] under different climate types and weather conditions for five-year period (from 2012 to 2016). Studied organic cotton fiber types were planted and cultivated in 2012, 2013, 2015 and 2016. Cotton fibers were not planted due to alternation and rotation in 2014. As earlier mentioned, Cumhuriyet 75 wheat variety was utilized as a rotation plant during the cultivation. Weather conditions for two different regions (Menemen and Sarayköy) over the studied years of 2012, 2013, 2015, and 2016 are given in Table 2. Temperature average values in Menemen location are lower than those of Sarayköy location where relative humidity values were higher in average.

2.2 Fiber Properties Determination

Fineness (micronaire index), fiber length UHM (mm), and fiber strength (g/tex) properties of studied organic cotton fiber types are measured using USTER HVI measurement system in Cotton Research Institute in Nazilli. Besides five types of studied fiber groups, there are two more independent variables are designated which are locations (2 different locations) and four years of consecutive cotton cultivations. Two different plantation locations of Büyük Menderes (Sarayköy/Denizli) and Bakırçay (Menemen/İzmir) were used for plantation of each 5 organic cotton fiber types for four

Table 1 Detailed information of used Turkish organic cotton fiber types

Registered name of the organic cotton fiber	Nazilli 84 S	Aydın 110	Nazilli DT 15	Emirel	Akdemir
Visual appearance					
Color of the fiber [1]	White	White	Camel hair	Brown	Brown
Scientific species name [1]	Gossypium hirsutum L.	Gossypium hirsutum L.	Gossypium hirsutum L.	Gossypium hirsutum L.	Gossypium hirsutum L.
Fiber details [1]	Developed via selective breeding from a Nazilli 84 variety and registered by NCRI[a], Nazilli 84 variety is a crossbred of Carolina Queen (Gossypium hirsutum L.) and 153-F (Gossypium hirsutum L.)	Crossbred of Ege 69 (Gossypium hirsutum L.) X Delcerro (Gossypium hirsutum L.) fiber varieties and registered by NCRI[a]	Crossbred of Nazilli 87 (Gossypium hirsutum L.) and Devetüyü (Gossypium hirsutum L.) fiber varieties and registered by NCRI[a]	A genotype developed via selective breeding from Devetüyü (Gossypium hirsutum L.) fiber variety which is originally supplied from USA GenBank and registered by EUAF&ÖVS[b]	A genotype developed via selective breeding from Devetüyü (Gossypium hirsutum L.) fiber variety which is originally supplied from USA GenBank and registered by EUAF&ÖVS[b]
Registered fiber length (UHM) (mm) [1]	28.5–29.5	34.1	26.5–27.1	25.2	28.0
Registered fiber fineness (micronaire) [1]	4.3–4.8	3.8–4.4	3.9–4.5	4.2	4.2–4.5
Registered average fiber strength (g/tex or lb/inch[b])[c] [1]	78,000–84,000 (lb/inch[b])	103,000–110,200 lb/inch[b]	21.7–25.7 g/tex	23.9 g/tex	27.2 g/tex

(continued)

Table 1 (continued)

Registered name of the organic cotton fiber	Nazilli 84 S	Aydın 110	Nazilli DT 15	Emirel	Akdemir
Registered ginning efficiency (%) [1]	44-45	34.3	33-39	36.8	34.5
Average days of maturity (days) [1]	120	117	119	109	113
Plant type [1]	In conical form	In semi-cylindrical form	In conical form	In conical form	In conical form

[a]NCRI: Nazilli Cotton Research Institute, Aydın, Turkey
[b]EUAF&ÖVS; Ege University Agricultural Faculty and Ödemiş Vocational School, İzmir, Turkey
[c] Average fiber strength of both white cotton fibers were registered in lb/inch^2 unit and that of three naturally colored cotton fibers were registered in g/tex unit

Table 2 Weather conditions of organic cotton fiber cultivation seasons according to their locations

Cultivation location	Month of the year	2012–2016 temperature (°C)			2012–2016 rain (average)		
		Max.	Min.	Average	Relative humidity (%)	Number of rainy days (Days)	Total rain (mm)
Menemen/İZMİR	May	27.1	14.3	20.5	61.0	6.5	46.0
	June	32.0	18.6	25.4	55.2	3.5	36.3
	July	34.7	21.8	28.4	46.1	0.3	0.4
	August	34.6	21.7	27.8	49.3	0.0	0.0
	September	30.6	17.2	23.4	58.3	0.8	6.4
	October	25.1	13.3	18.4	66.2	3.8	46.0
Sarayköy/DENİZLİ	May	27.2	14.8	20.5	50.5	10.5	41.9
	June	33.3	18.9	26.0	40.3	4.8	18.8
	July	36.6	22.2	29.2	34.5	2.8	5.1
	August	36.0	22.0	28.3	34.9	3.0	10.4
	September	32.1	17.8	24.3	42.8	3.3	13.3
	October	25.4	12.7	18.2	53.3	4.3	27.2

consecutive years of fiber crop. Independent variables of 5 fiber types, 2 locations, and 4 cultivation years (2012, 2013, 2015, and 2016) were designated to manage statistical evaluation on dependent fiber variables of thickness/fineness (micronaire), fiber length (mm), and fiber strength (gr/tex). All fiber property measurements are conducted in the fiber measurement laboratory of Nazilli Cotton Research Institute on a High Volume Instrumentation—HVI system [6].

3 Results and Discussion

3.1 Cotton Fiber Properties Comparison

Fiber length is one of the major quality issues on the cotton fiber evaluation. Naturally colored organic cotton fibers are generally known with their short fiber length comparing to the common white cotton fiber types. In Fig. 1, length change of the each five fiber types is exhibited depending on the location and crop year. As it can be seen, the highest fiber length belongs to the white fiber type of Aydın 110 at each growing conditions. The lowest fiber length property belongs to the colored fiber type of Akdemir. Crop years of 2012, 2013, 2015, and 2016 seem to be as an influential factor on the length of each cotton fiber type (Fig. 1). Location of the cultivation land seems having some influences depending on the weather conditions in detail, however trends based on fiber type and crop year seems similar, since both locations are in the Aegean region of Turkey.

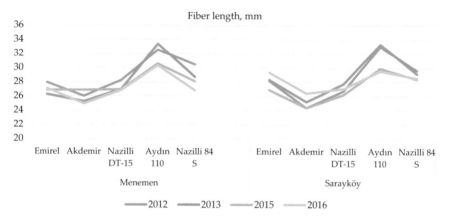

Fig. 1 Length change of the cotton fiber samples depending on the cotton fiber type, year and location of growth

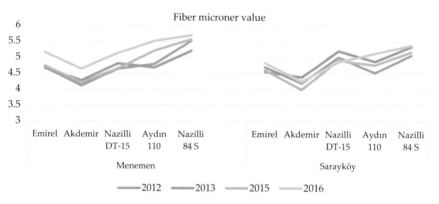

Fig. 2 Fineness change of the cotton fiber samples depending on the cotton fiber type, year and location of growth

Fiber fineness is another important quality issue on the cotton fiber evaluation. Naturally colored cotton fibers are generally known with their thinner fiber form with lower micronaire value comparing to the common white cotton fiber types. In Fig. 2, fineness change of the each five cotton fiber types is shown depending on the location and crop year. As it is seen, the highest fiber fineness value belongs to the white fiber type of Nazilli 84 S at each growing conditions. The lowest fiber fineness property belongs to the colored cotton fiber type of Akdemir. Crop years of 2012, 2013, 2015, and 2016 seem to be as an influential factor on the fineness of each cotton fiber type (Fig. 2). Location of the cultivation land seems having some influences depending on weather conditions in detail.

Fiber strength is also important quality issue on the cotton fiber and yarn literally. Naturally colored cotton fibers are generally known with their lower fiber strength

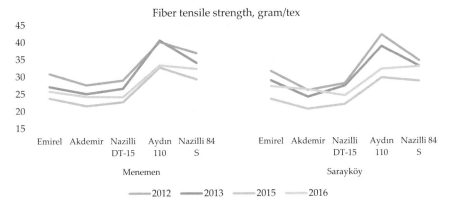

Fig. 3 Strength change of the cotton fiber samples depending on the cotton fiber type, year and location of growth

value comparing to the common white cotton fiber types. In Fig. 3, tensile strength change of the each five fiber type is shown depending on the location and crop year. As it is seen, the highest fiber strength value belongs to the white fiber type of Aydın 110 at each growing conditions. The lowest fiber strength values belong to the colored fiber types of Akdemir and Nazilli DT 15. Crop years of 2012, 2013, 2015, and 2016 seem to be as an influential factor on the strength property of each fiber type (Fig. 3). Location of the cultivation land seems having some influences depending on weather conditions in detail.

3.2 Statistical Evaluation of Cotton Fiber Properties

Statistical evaluation has been designed to use Least Squares Fit model to accomplish Analysis of Variance and Effect Tests of each individual dependent variable of fiber type, location, year, and their interactions of fiber type*location, fiber type*year, location*year, and fiber type*year*location.

Independent variables of fiber length, fiber fineness, and fiber strength are evaluated using above defined statistical approaches at α significance level of 0,05 and designated F values are shown in Table 3.

Fiber length

It is seen on Table 3 that fiber length is influenced by the dependent variable of year, fiber type, interactions of location*year, fiber type*year, and fiber type*location*year. It can be stated that fiber length of each fiber type is influenced by the year and location differences of weather differences of temperature, humidity, and rain.

In Table 4 statistical evaluation results for length property of the fiber samples depending on the type, year, and location are shown in detail.

Table 3 Significance level of dependent variables on independent variables

Source	Prob>F		
	Fiber length (mm)	Fiber fineness (mic.)	Fiber strength (g/tex)
Year	<0.0001	<0.0001	<0.0001
Location*year	0.0002	<0.0001	0.1589
Fiber type	<0.0001	<0.0001	<0.0001
Fiber type*year	<0.0001	0.0486	<0.0001
Fiber type*location*year	0.0063	0.0535	0.1174

Table 4 Level test plots and classification for fiber length (mm)

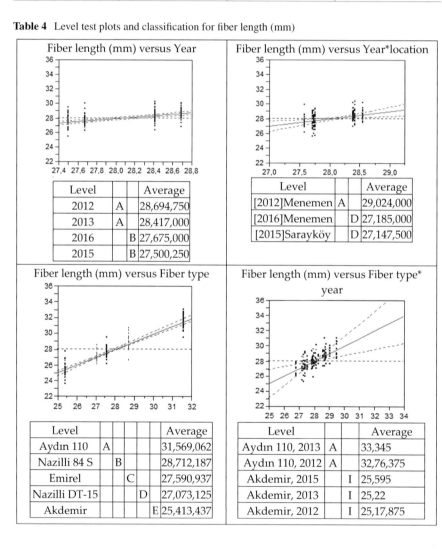

Fiber length (mm) versus Year

Level		Average
2012	A	28,694,750
2013	A	28,417,000
2016	B	27,675,000
2015	B	27,500,250

Fiber length (mm) versus Year*location

Level		Average
[2012]Menemen	A	29,024,000
[2016]Menemen	D	27,185,000
[2015]Sarayköy	D	27,147,500

Fiber length (mm) versus Fiber type

Level					Average
Aydın 110	A				31,569,062
Nazilli 84 S		B			28,712,187
Emirel			C		27,590,937
Nazilli DT-15				D	27,073,125
Akdemir				E	25,413,437

Fiber length (mm) versus Fiber type* year

Level		Average
Aydın 110, 2013	A	33,345
Aydın 110, 2012	A	32,76,375
Akdemir, 2015	I	25,595
Akdemir, 2013	I	25,22
Akdemir, 2012	I	25,17,875

Fiber length versus year: The highest fiber length value belongs to the fibers of year 2012 (286,947 mm), and the lowest value (275,002 mm) belongs to the fibers of year 2015; fibers of 2012 and 2013 and fibers of 2016 and 2015 are grouped under two groups of A and B.

Fiber length versus year*location: The highest fiber length value (29,024 mm) belongs to the fibers of year*location of 2012*Menemen, and the lowest value (271,475 mm) belongs to the fibers of year 2015*Sarayköy.

Fiber length versus fiber type: The highest fiber length value (315,690 mm) belongs to the fiber type of Aydın 110, and the lowest value (254,134 mm) belongs to the fiber type of Akdemir. Each five type of fiber length is found significantly different then each other as they are grouped under five different letters of A, B, C, D, and E.

Fiber length versus Fiber type*year: The highest fiber length value (33,345 mm) belongs to the fiber type of Aydın 110*2013, and the lowest value (2,517,875 mm) belongs to the fiber type of Akdemir*2012. Fiber length difference against interaction of fiber type*year is levelled under only two letters of A and D where 2012*Menemen is A and 2016 Menemen and 2015 Sarayköy are D.

Fiber Fineness

It is seen on Table 3 that fiber fineness is influenced by the dependent variable of year, fiber type, interactions of location*year, and fiber type*year. Fiber type*location*year interaction is not found significant. It can be stated that fineness of each cotton fiber type is influenced by the year and location differences of weather differences of temperature, humidity, and rain. Especially crop year is found the most influential factor on the fiber fineness.

In Table 5 statistical evaluation results for fineness property of the fiber samples depending on the type, year, and locations are shown in detail.

Fiber fineness versus year: The highest fiber fineness value belongs to the fibers of year 2016 (50,522,500), and the lowest value (47,180,000) belongs to the fibers of year 2012; fibers of 2012, 2015 and 2013 and fibers of 2016 are grouped under two groups of A and B.

Fiber fineness versus year*location: The highest fiber fineness value (52,195,000) belongs to the fibers of year*location of 2016*Menemen, and the lowest value (46,850,000) belongs to the fibers of year 2015*Sarayköy.

Fiber fineness versus fiber type: The highest fiber fineness value (53,665,625) belongs to the fiber type of Nazilli 84 S, and the lowest value (42,450,000) belongs to the fiber type of Akdemir. Fiber fineness of fiber types are found significantly different than each other, except fiber types of Aydın 110 and Nazilli DT-115 which are grouped together under B.

Fiber fineness versus Fiber type*year: The highest fiber fineness value (55,325,000) belongs to the fiber type of Nazilli 84 S*2016, and the lowest value (40,887,500) belongs to the fiber type of Akdemir*2015.

Fiber Strength

It is seen on Table 3 that fiber strength is influenced by the dependent variable of year, fiber type, and interactions of year and fiber type. Location*year, and Fiber

Table 5 Level test plots and classification for fiber fineness (micronaire)

Fiber fineness versus year	Fiber fineness versus location and year

Level		Average
2016	A	50,522,500
2013	B	47,917,500
2015	B	47,765,000
2012	B	47,180,000

Level		Average
[2016]Menemen	A	52,195,000
[2012]Sarayköy	D	46,945,000
[2015]Sarayköy	D	46,850,000

Fiber fineness versus fiber type	Fiber fineness versus fiber type *year

Level			Average
Nazilli 84 S	A		53,665,625
Aydın 110		B	49,318,750
Nazilli DT-15		B	48,965,625
mirel		C	47,331,250
Akdemir		D	42,450,000

Level		Average
Nazilli 84 S, 2016	A	55,325,000
Akdemir, 2012	G	41,400,000
Akdemir, 2015	G	40,887,500

type*location*year interaction are not found significantly influential on the fiber strength value. In Table 6, statistical evaluation results for strength property of the fiber samples depending on the type, year, and location are shown in detail.

Fiber strength versus year: The highest fiber strength value belongs to the fibers of year 2012 (32,682,500), and the lowest value (25,460,000) belongs to the fibers of year 2015. Each four years of fiber strength is grouped under four different groups implying the significantly different on fiber strength values of each different crop year.

Fiber strength versus year*location: The highest fiber strength value (32,865,000) belongs to the fibers of year*location of 2012*Menemen, and the lowest value (24,920,000) belongs to the fibers of year 2015*Sarayköy.

Table 6 Level test plots and classification for fiber strength (g/tex)

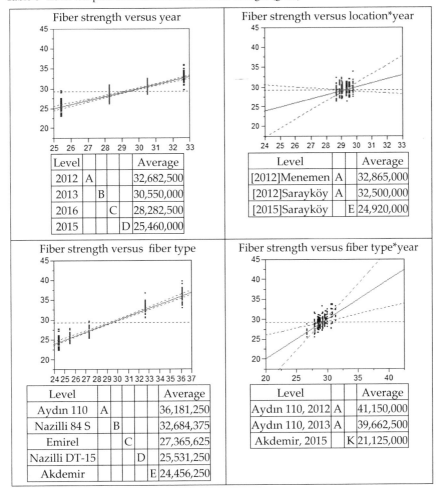

Fiber strength versus year				
Level				Average
2012	A			32,682,500
2013		B		30,550,000
2016			C	28,282,500
2015			D	25,460,000

Fiber strength versus location*year			
Level			Average
[2012]Menemen	A		32,865,000
[2012]Sarayköy	A		32,500,000
[2015]Sarayköy		E	24,920,000

Fiber strength versus fiber type						Average
Aydın 110	A					36,181,250
Nazilli 84 S		B				32,684,375
Emirel			C			27,365,625
Nazilli DT-15				D		25,531,250
Akdemir					E	24,456,250

Fiber strength versus fiber type*year			Average
Aydın 110, 2012	A		41,150,000
Aydın 110, 2013	A		39,662,500
Akdemir, 2015		K	21,125,000

Fiber strength versus fiber type: The highest fiber strength value (36,181,250) belongs to the fiber type of Aydın 110, and the lowest value (24,456,250) belongs to the fiber type of Akdemir. Each five type of fiber strength is found significantly different than each other.

Fiber strength versus Fiber type*year: The highest fiber fineness value (41,150,000) belongs to the fiber type of Aydın 110*2012, and the lowest value (21,125,000) belongs to the fiber type of Akdemir*2015.

In Table 6 statistical evaluation results for tensile strength property of the cotton fiber samples depending on the type, year, and locations are shown in detail. It can be stated that strength of each cotton fiber type is influenced by the year and location differences (temperature, humidity, and rain). Climate differences of different years are found the most influential factor on the fiber strength.

4 Conclusions

Cotton fiber cultivation and its processing in the modern textile manufacturing lines are major agricultural and industrial activities in Turkey. Leading researches about cotton fiber types are continued in Turkey and in the world in order to increase the cotton yield per acre and moreover to improve fiber properties and to adapt fiber types to the different type of soil and weather conditions. Different weather conditions on the cultivation location could affect the final fiber properties. Cotton fiber breeders are also working to improve the properties and quality of different types of cotton fiber genotypes (white, off-white and naturally colored) by using the conventional plant breeding techniques such as a selection, crossing and mutation and utilizing biotechnology leading to the production of better and more productive superior varieties of cotton fibers.

In this study, white and naturally colored unique Turkish organic cotton fibers which were developed via crossbreeding and selective breeding techniques in Turkey are analyzed depending on the physical fiber properties. The physical properties such as fineness (micronaire index), fiber length, and fiber strength of different varieties of two white (Nazilli 84 S and Aydın 110) and three naturally colored (Emirel, Akdemir, Nazilli DT-15) unique Turkish organic cotton fiber types which cultivated in two different plantation locations in the Aegean Region of Turkey [Menemen/İzmir (Bakırçay Basin) and Sarayköy/Denizli (Büyük Menderes Basin) locations] under different climate types and weather conditions for five-year period (from 2012 to 2016) were investigated and discussed with the statistical evaluation using Least Squares Fit model to accomplish Analysis of Variance and Effect Tests. Since, fiber properties of length, strength and fineness are the most influential cotton fiber properties on the yarn processing efficiency and yarn properties. Different type of cotton fibers, different weather conditions of crop year and different cultivation location are found somehow influential factors on the studied major fiber properties. Statistical evaluation of fiber length has shown that length changes depending on the crop year and location differences where seasonal weather conditions vary. In average values, it is seen that the highest fiber length belongs to the white fiber type of Aydın 110 at each growing condition. The lowest fiber length property belongs to the naturally colored fiber type of Akdemir naturally colored cotton fiber. Statistical evaluation of fiber fineness has shown that fineness changes depending on the crop year, fiber type, interactions of location*year, and fiber type*year. Especially crop year is found the most influential factor on the fiber fineness, that the same fiber exhibits different fiber fineness values depending on the slight weather condition changes of year and also location. In average values, it is seen that the highest fiber fineness value belongs to the white fiber type of Nazilli 84 S at each growing condition. The lowest fiber fineness property belongs to the naturally colored fiber type of Akdemir. Statistical evaluation of fiber strength of each fiber type is influenced by the crop year and location differences. Seasonal climate differences of different years are found the most influential factor on the fiber strength. In average values, it is seen that the highest fiber strength value belongs to the white fiber type of Aydın 110 at each growing condition. The lowest fiber strength property belongs to the colored organic cotton fiber type of Akdemir and Nazilli DT 15. Cotton fiber type of Aydın 110, which is white cotton fiber, is found having the best fiber properties among the five fiber type, and naturally colored organic cotton fiber type of Akdemir, which is one of the studied naturally colored organic cotton fiber, is found having the lowest fiber properties in general.

Organic cotton and naturally colored cotton growing in the World has found a widening ground in the textile and clothing markets that has environment sensitive consumers. Organic cotton and also naturally colored cotton have relatively less desired fiber properties comparing to commonly grown well engineered modern cotton types. Environmental aware consumer market size in textile and clothing sector, changing fashion trends are two of main influential issue on the size of organic cotton and naturally colored cotton market. Some of the world wide major textile and clothing brands are important players in the World organic cotton market. Turkish organic cotton market and naturally colored cotton market will continue to be part of this market. There are well trained farmers, private companies and government initiations in Turkey to help building of permanent organic cotton growing capacity. Global textile and clothing market conditions and restrictions are the most influential factor on the feature of organic cotton growing in Turkey. Studies on the analyzing and improvement of fiber properties will continue in local cotton growing farms, cotton research institutes, and yarn—textile manufacturers in Turkey.

References

1. Republic of Turkey Ministry of Customs and Trade, Directorate General of Cooperatives (2016) Cotton Report. Republic of Turkey Ministry of Customs and Trade, Ankara
2. ICAC (2017) https://icac.gen10.net/statistics/index2017. [Online]. Accessed on 08 July 2017
3. Devrent N, Palamutcu S (2017) Mini review on organic cotton. J Text Eng Fashion Technol 3(2)
4. Unsal A (2017) https://www.icac.org/tis/regional_networks/documents/asian/papers/unsal.pdf. [Online]. Accessed on 30 Aug 2017
5. Better cotton initiative 2016 Annual Report. Genève—Switzerland
6. Uster Technologies (2011) https://www.uster.com/en/instruments/fiber-testing/uster-hvi/. [Online]. Accessed on 05 Sept 2017

Sustainability Goes Far Beyond "Organic Cotton." Analysis of Six Signature Clothing Brands

María Lourdes Delgado Luque and Miguel Angel Gardetti

Abstract In line with the criteria described in the Commission of the European Communities, Commission Recommendation of 6 May 2003, the concept referred to as "signature clothing design" is called "microenterprise" in Spain. Based on such criteria, business organisations which annual turnover does not exceed €2 million and have less than 10 employees fall into this category. This segment is part of the textile and clothing value chain in Spain, and it stands out for its innovation and originality, both in terms of the end product and the different stages comprising the project and production process. The creations of microenterprises do not follow the trends set by large fashion centres; on the contrary, they produce goods that convey a specific identity that feeds on the geographic, production and cultural environment. Many brands claim that they are "sustainable" when it comes to communications, customer relationship, and product sales. While such claim is mainly based on materials—use of certified organic cotton—it also includes other aspects, for example, the use of natural dyes. Moreover, many of these brands are referred to as sustainable by different Spanish organisations that have set their own sustainable classification criteria, such as Slow Fashion Next, Moda Sostenible de Zaragoza, Asociación Moda Sostenible de España, Slow Fashion Aragón, etc. This chapter analyses five Spanish brands based on the sustainability criteria defined by the authors. For such purpose, we will analyse all the public information referred to by the brands: websites, newschapter articles, references from organisations, and case studies, if any. We will also interview each of the designers or owners of these microenterprises. This will be compared to a model developed by the authors that addresses the meaning of being sustainable in the textile and fashion world. The final chapter will include conclusions and recommendations.

Keywords Sustainable fashion · Spain · Organic cotton

M. L. Delgado Luque (✉) · M. A. Gardetti
Sustainable Textile Center, Imperio Argentina 12, Gate 2, 4th B, 29004 Malaga, Spain
e-mail: ml@ctextilsustentable.org.ar

M. A. Gardetti
Center for Studies on Sustainable Luxury, Paroissien 2680, 5th "B",
C1429CXP Buenos Aires, Argentina

© Springer Nature Singapore Pte Ltd. 2019
M. A. Gardetti and S. S. Muthu (eds.), *Organic Cotton*, Textile Science
and Clothing Technology, https://doi.org/10.1007/978-981-10-8782-0_3

1 Introduction

The term "sostenibilidad" (sustainability, in English) was added to the dictionary of the Royal Spanish Academy (RAE) only in 2010. This definition only refers to the environment and the natural resources, as opposed to the conviction established in the institutional and business arenas that sustainability also includes the social aspect [1] and the individual per se [2]. The purpose of this chapter is to prove that, as illustrated by the dictionary of the Royal Spanish Academy, brands are still laying emphasis on raw materials and some specific aspects of the environmental side when it comes to defining themselves as "sustainable," downplaying the other concepts that are part of the general sustainability[1] picture.

Along this line, Olga Algayerova, Executive Secretary of the United Nations Economic Commission for Europe, describes the situation caused by the fashion industry as an "emergency".[2] This expression becomes highly visual. This is worrying not only because of its environmental consequences, but also because the term "emergency" can be applied to the disasters caused by overconsumption in fashion [3, 4].

In Spain, the development of sustainable (sustainable?)[3] fashion was mainly led by a large number of entrepreneurs. In line with the criteria described by the European Commission in Commission Recommendation dated 6 May, 2003, the analysed brands are classified as "microenterprises" based on their business characteristics, number of employees (less than 10), and annual turnover, which should be less than 2 million euro.

In Spain, this segment—which is characterised by its innovation and originality—has loomed large in the textile and clothing value chain, both in terms of the end product and in the different stages of the project and the production process. Moreover, the creations of microenterprises do not follow the trends set by large fashion centres; on the contrary, they produce goods that convey a specific identity that feeds on the geographic, production and cultural environment.

Brands that fit into the above segment claim that they are "sustainable" in their communications, customer relationships, and product sales. While this claim is mainly based on materials—the most important being certified organic cotton—in some cases, it also includes other aspects, such as the use of natural dyes.

Likewise, many of these brands are classified as sustainable by different Spanish organisations which have set their own sustainable classification criteria, such as Asociación Moda Sostenible de España (AMSE), Asociación de Moda Sostenible de Barcelona (AMSB), Moda Sostenible de Zaragoza (Catálogo Deseclipsando), Slow Fashion Next (Moda en Positivo), etc.

This research study addresses how the brands that claim to be sustainable turn to organic cotton as the main argument for the communications speech. The method-

---

[1] Also referred to as "big picture" by Gardetti [2].

[2] Olga Algayerova, was part of a High-Level Panel titled "Fashion and the Sustainable Development Goals: What role for the UN?," where she told the delegates that "promoting sustainable consumption becomes an imperative to address the many emergencies created by fashion" [3].

[3] Authors ask this, rhetorically.

ology development deals with the guidelines used for this research study. The third section explains the meaning of sustainable in the textile and fashion sector based on the model proposed by Gardetti [2], which will help analyse the data from the brands selected for this study. Finally, the conclusions are described in the last section.

2 Methodology

In addition to conducting a documentary search for theoretical grounds, a qualitative methodology for data collection and exploitation is used in this chapter.

Data were collected based on documentary searches of published and free-access information (websites, newschapter articles, references from organisations, and case studies) about the brands.

Six Spanish brands—referred to as Brand 1, Brand 2, Brand 3, Brand 4, Brand 5 and Brand 6—were analysed based on the following criteria defined by the authors: business typology homogeneity in terms of the number of employees, turnover, nationality, participation in a sustainable fashion platform, and information accessibility. And they are jointly representative of sustainable fashion in Spain.

The technique of semi-structured qualitative interviews with open answers [5] was also used to get information straight from the sources. The benchmark for the questions was the model proposed by Gardetti [2] about the meaning of being sustainable in textiles and fashion. The questions are listed in Appendix.

One representative of each sustainable brand was interviewed. Interviews were e-mailed to brand leaders. All interviews were answered within a short period of time. Table 1 shows each interviewee's academic degree and profile, and each interview date.

Brand leaders provided the information on condition that it should be exclusively used for this academic study in order to prevent information disclosure or provision to third parties that are not involved in the study along with their brand name. For such purpose, every piece of information which might lead to brand identification

Table 1 Each interviewee's academic degree and profile and each interview date

Representative	Academic degree	Profile	Interview date
Brand 1	Architect	Entrepreneur	06/03/2018
Brand 2	Graphic designer	Entrepreneur	26/02/2018
Brand 3	Marketing	Entrepreneur	31/01/2018
Brand 4	Design	Entrepreneur	06/03/2018
Brand 5	Design and pattern	Entrepreneur	02/03/2018
Brand 6	English philology	Entrepreneur	10/03/2018

Source Prepared by the authors

was removed. Collected data were compared to the model developed by Gardetti [2]. Finally, the conclusions and recommendations are analysed.

While all the brands carefully selected for this study are representative of sustainable fashion in Spain, this study is limited regarding the number of brands under review.

3   What Does It Mean to Be Sustainable in the Textile and Fashion Sector?

An analysis published by the prestigious Spanish magazine Modaes.es about the sustainable fashion outlook in Spain argues that the leading industry brands are taking action in this field, not so much to create sustainable fashion, but to turn processes sustainable [6]. However, in the signature brand or microenterprise sector, the vocation of the so-called "sustainable natives" does not believe in separating product from production processes [2].

The definition issued by the European Commission includes environmental and social aspects, but with no restriction as to types of industries, business model or size; hence, both agencies and institutions made different proposals to narrow the concept down [1, 2, 7, 8].

The commercial aspect has played an important role in this initiative to create sustainable fashion, which is inconsistent with the reduced consumption proposed by sustainability. As shown by reality, fashion brands do not produce necessity goods, [1, 9] which means that no "fashion" garment is inherently sustainable.[4]

Experts argue that today, brand is still regarded as an added value, but in the near future, it will become essential for brand viability [1, 2, 4]. An example of the above is that there is still a long way between the statement of principles made by companies in their sustainability reports and their actual business practices [1].

In line with the subject matter of this research study, it is worth pointing out the Spanish consumers' sensitivity in such connection. That is, what do the Spanish people think about responsible consumption and sustainable development? Based on the report prepared by Club de Excelencia en Sostenibilidad—with the assistance of consulting firm Nielsen—the general opinion of the Spanish people about sustainability is related to environmental issues. The Spanish people believe that economic interests are the main barrier to sustainability. In general, they claim that brands offer little information about their actions for sustainability and they are convinced that regulation, education and information are key to promoting responsible consumption [10].

Based on the **model** proposed by **Gardetti** [2], sustainable reality underlines economic, environmental and social aspects—all with a strong dynamic, interactive, and multidisciplinary nature—which reveal the systemic aspect of sustainability. That is, to understand sustainability, it is essential to open your mind to the "**big picture**".

---

[4][1, 2, 5–7, 9, 10].

Fig. 1 Aspects related to sustainable fashion and textiles. *Source* Gardetti [2]. Published with Editorial LID Argentina's authorisation

This means to recognise that sustainability is about well-being and the long term, while raising awareness to accurately reflect its implications, such as a keen vision of reality, its protection, and our responsibility not to delay necessary changes. Figure 1 shows an overview of the aspects that have a direct impact on sustainability—environmental aspects, consumers, design and innovation [17]—as well as technological and social aspects, business models, marketing and communications, and raw materials.

An approach based on the "big picture" will help identify such aspects with an impact on fashion and textiles sustainability, including economic sustainability, as assurance of ongoing and sufficient cash flow to ensure readily available funds while getting returns that provide for brand viability. Figure 2 shows all the aspects related to sustainable fashion and textiles: the "big picture".

Environmental aspects range from the efficient use of water—including water pollution and waste water—energy and soil, to gas emissions—CO_2 and other air emissions—in addition to processes related to hazardous materials and substances—integrating human toxicity—and waste management throughout their entire cycle, including disposal.

On the one hand, the consumer figure involves access to information and responsible purchase decisions. On the other, it shows their responsibility in terms of clothing care: efficient (or inefficient) use of water and energy, use of cleaning chemicals—with no disregard for consumer decisions about what to do with the clothes that they no longer wear (waste/donations).

Innovation, understood as "incremental continuous improvement," has been overdeveloped in our global reality. Most likely, the best question that we should ask ourselves is what type of innovation we need to achieve sustainability, and the answer will lead us to "disruptive innovation" [2], that provides for a systemic transformation. The advent of the new sustainable paradigm [11] is making brand reaction possible, increasing their commitment to products, processes and the "existing" markets [2].

The technological aspects include product-related processes and issues, while the social aspects involve labour rights, human rights, social inclusion, human trade, modelling, anti-corruption and fast fashion knowledge and understanding.

Moreover, the business world has no problem to recognise that sustainability is part of the business model and a condition to conduct business [2]. In turn, marketing

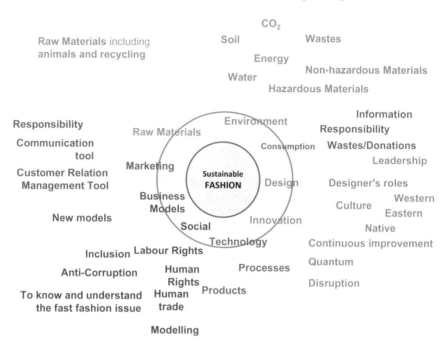

Fig. 2 All the aspects related to sustainable fashion and textiles: the "big picture". *Source* Gardetti [2]. Published with Editorial LID Argentina's authorisation

becomes a tool to communicate the sustainable message to customers. Therefore, marketing is a great ally of branding, since sustainability is monopolising part of its speech.

Finally, raw materials include not only raw material in its broad sense, but also issues related to the use of water, energy, and soil; and aspects related to animals and recycling. The relevance taken on by techniques of redesigning, cutting and making from scratch an entire new garment or parts of it from old pieces, fabrics (vintage) and accessories has brought about a major transformation in the raw material universe. Therefore, the value of used materials has changed. Conversely, in downcycling, the value of the new material is lower than that of the original item. In this case, innovation and the emergence of new materials from recycling will gradually reduce downcycling deficits.

According to Gardetti [2], systemic vision and transparency entail two aspects projected across the entire value chain: company raw materials/start-up/supply chain. The importance of transparency is particularly evidenced in supply chain tracking and monitoring (see Figs. 3 and 4) per se, which is essential for stakeholder sustainable management. Therefore, this value should be prioritised, as it goes from sourcing the raw material to the retail business; the supply chain impacts differently on stakeholders, and the end product affects both the environment and the society.

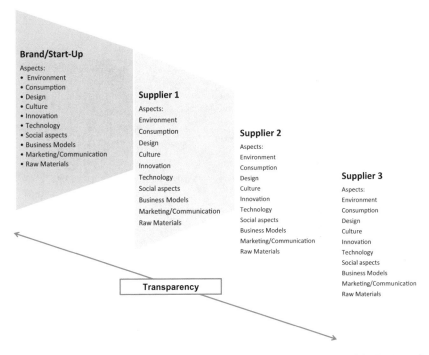

Fig. 3 Supply chain transparency. *Source* Gardetti [2]. Published with Editorial LID Argentina's authorisation

The systemic nature of sustainability reveals how every aspect involved relates with one another. Thus, the "big picture" shows that carbon dioxide emissions—mainly coming from clothes and textile manufacturing and care processes—are closely related to climate change, which, in turn, affects raw material production processes, mainly cattle and farm; or the efficient or inefficient use of water and energy, as regards extraction and production of raw materials; as well as everything related to consumer information access and responsibility (information on the use of these resources in the production process and responsibility in taking care of garments). Moreover, waste (mainly textile waste) has an impact on both materials and the recycling process [22]. In turn, raw materials are closely related to garment "traceability," i.e., the possibility to trace back the full product information based on a unique ID, all the way from component sourcing, through each stage of the supply chain process, up to its final disposal. There is no doubt that consumer sensitivity is increasingly interested in the purchase decision information. This also includes ethical aspects related to raw material production, product manufacture, distribution and sale. In other words, the aspects included in Fig. 5 should be considered throughout the chain and from the beginning, paying special attention to stakeholders.

Finally, the cultural perspective is not left outside the different processes, including the purchase decision and the role played by consumers. In this connection, designers

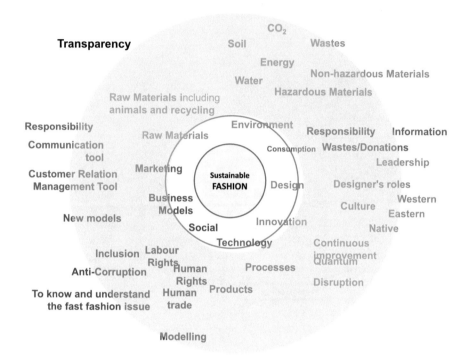

Fig. 4 Relationship between the aspects involved in sustainable fashion and transparency. *Source* Gardetti [2]. Published with Editorial LID Argentina's authorisation

play a leading role in the reappraisal of native culture and the consideration of other cultures.

Along with the above, the most relevant aspect of the model proposed by Gardetti [2] is to consider textiles and fashion sustainability at an individual level. For such purpose, we should carefully think about who we are and how we manage the world using science and technology, with a particular focus on the humanistic and ethical domains in terms of being accountable for our own actions.

4   Certifications

In view of the difficulty to define what sustainability in the textile and fashion sector is all about, and how to measure it, some proposals in the form of seals, labels, and certifications have emerged [1, 12]. In general, the textile certifying process has different dimensions: on the one hand, raw material certification and, on the other, the certification of the textiles manufactured with them. In this case, the substances used during manufacture—such as dyes and trimmings—and product traceability should be taken into account, along with raw materials.

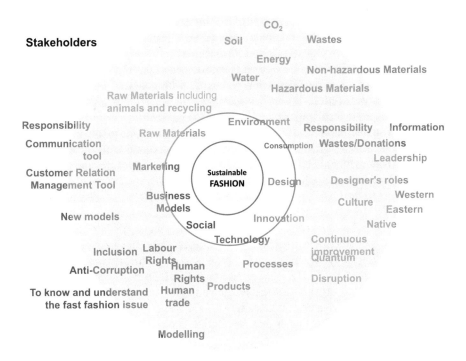

Fig. 5 Aspects related to sustainable fashion and textiles and stakeholders. *Source* Gardetti [2]. Published with Editorial LID Argentina's authorisation

In addition to seals, labels, and certifications, companies have another practice that favours sustainable principle monitoring and verification: the so-called Sustainability Reports. These reports are prepared based on guidelines set by the Global Resource Initiative (GRI). This document is the result of a unique consulting participatory process which includes representatives from reporting organisations and users of these reports all over the world. The guide to prepare sustainability reports is designed so that every (large and small) organisation in the world can use it—both experienced report writers and people who come into contact with sustainability reports for the first time [8].

In line with the above, the absence of official regulations has favoured the emergence of private certifications. While governments have passed laws on labour and environmental issues, when it comes to sustainability there are factors which either fall outside of such regulations or become minimum requirements.[5]

This study will briefly address the certifications held by the analysed brands, i.e., GOTS, OCCGUARANTEE and Standard 100 by OEKO-TEST.

---

[5]The European legislation reserves the term bio for certified, ecologically grown agricultural products (Regulation (EEC) No. 2092/91). This regulation was supplemented by other regulations to include the terms biological, eco-, ecological, organic, and biodynamic (EC No. 1804/99). Spain implemented this regulation in July, 2006 [13].

The GOTS certification is comprised of four member organisations: Organic Trade Association (OTA), IVN (International Association of Natural Textile Industry), Soil Association, and Japan Organic Cotton Association.

The GOTS[6] standard establishes a set of requirements that companies should meet during the knitting and manufacturing processes, with a special focus on organic cotton. In this connection, it certifies that the cotton used was grown free from pesticides and toxic fertilizers, and without transgenic seeds.

GOTS is an international standard created by members of the textile industry and different organisations with textiles standards, which also counts with the cooperation of IFOAM (International Federation of Organic Agriculture Movements) in order to set harmonised criteria applicable worldwide.

According to the GOTS public database, Spain has 1050 brands certified by this entity. The GOTS system is based on inspections and certification of processors, manufacturers and traders performed by independent and GOTS-accredited bodies in order to provide a credible assurance for the integrity of textile products. To protect GOTS' credibility, they investigate and impose sanctions wherever there is evidence of misleading use of the GOTS label or reference to GOTS certification.[7]

Another certifying organisation that has gained a great number of members among fashion brands is the organic cotton stamp OCCGUARANTEE, created by Organic Cotton Colours,[8] a supplier that, in turn, has GOTS (Global Organic Textile Standard) certification with CU832882.

They produce 100% organic cotton, with colours born from the Earth and zero dyes, which farming has a great social impact on the 450 families that are part of the Organic Colours Cotton project. The project representative states that "it represents our firm commitment to caring as much as possible for people, the environment and all the farmers involved in growing and producing the organic cotton used by Organic Cotton Colours for 25 years".[9]

The Organic Cotton Colours company was founded by Ángel Sánchez in the 1990s. In 2010, Santi Mallorquí, current CEO, decided to continue the project with the same values and philosophy, but adding a new social dimension: to start working along with farmers from Brazil, a country that meets the requirements to be able to continue growing colour cotton.

The Brazil growing project is possible thanks to all the famers and families with whom they work as a team. "Our goal is to generate a real social impact: to help all families live on their own crops and make profit from cotton".[10]

---

[6]The documentary references about the GOTS certification were taken from http://www.global-standard.org/es/, and from other online sources referenced in the text.

[7]Information updated on 9 December, 2017, at 12:54 am, on http://www.global-standard.org/es/protect.html.

[8]The information about OCCGUARANTEE was taken from its website www.organiccolours.com and discussed with Santi Mallorquí, brand leader back when this chapter was prepared. The authors appreciate his kind cooperation.

[9]Ibid.

[10]Ibid.

Finally, in connection with dye certification, we can mention the Swiss group OEKO-TEX, specialised in developing standards for the textile industry and issuer of two certifications: Standard 100 by Oeko-Text, to detect harmful substances, and Standard 1000 by Oeko-Tex, for environmental management.

The origin of these two standards can be traced back to the AITEX research institute (Valencia), which created the Made in Green label in 2005 [14]. This label ensures that brands that used it have a sustainable production chain [15].

Since its introduction in 1992, the central focus of the standard has been the development of test criteria and limit values, on the basis of its comprehensive and strict catalogue of measures, with several hundred regulated individual substances which, in many cases, go far beyond the applicable national and international standards. The OEKO-TEX® tests for harmful substances are mainly based on the purpose of textiles and materials [14, 16].

4.1 Certifications as Sustainability Paradigm?

The proliferation of certifications becomes—at least—a warning about the means implemented to do things properly. The mere fact that there are mandatory requirements in place means a lot. Of course, holding certification does not ensure that a product is partially or wholly sustainable. It just means that the brand has claimed to be sustainable in certain aspects.

In addition to the certification boom, it is evident that associations and platforms have become a shop window for brands and certifications alike. They have also embraced the need to define and ensure what sustainability is and is not. For such purpose, in most cases they look for brand certifications, as they have no other monitoring tool. In fact, brands authenticate themselves. An example of this is the case of the platform called Moda Impacto Positivo supported by Slow Fashion Next. The partnering and collaborative purpose that triggered the creation of this fashion platform is praiseworthy. The entity has defined some symbols[11] to help identify sustainability aspects that can be found in the brands. The accuracy of this data is declared by signing a free access document on the website.[12] Brands' self-declaration includes a disclaimer regarding its contents, holding both the platform and its founder harmless.

The operation of associations and platforms evidences that stamps and certifications about partial aspects of processes are serving as assurances that all the brand's processes are sustainable. One way or another, they are proving their ability to set and view metrics in a rather difficult field to achieve this goal.

---

[11] The symbols that identify the sustainable aspects are available on https://modaimpactopositivo.com/simbolos-consumo-sostenible/.

[12] The supporting document is available at the following link https://modaimpactopositivo.com/slowfashion-marcas-complementos-ropa-ecologica-online/. Click on the "See certificate" button to display one of the brands.

5 Some Organisations that Promote and Advertise "Sustainable Fashion"

The collaboration and cooperation concept is deeply rooted in sustainability. While the original goal was to join forces, the reality is quite different. The emergence of a great number of associations has caused fragmentation—which set forces apart, instead of joining them. This chapter will only address the role played by these platforms in offering a vision about those associations and relating them to the above brands.

The main goal of sustainability-promoting organisations and platforms is to offer visibility to entrepreneurs in favour of a meeting point for customers and suppliers. Most of them are made up of brands or people who—after years working at major companies in the fashion industry, or even coming from other sectors—have made a social and ethical commitment to social and environmental sustainability.

In Spain, while sustainable fashion platforms and organisations were conceived to join forces and spread the sustainable fashion concept, in reality, the commercial purposes have blurred the original goals.

Organisations are aware of this and, therefore, they refer to it on their social media, evidencing the disruptive environment in which they were immersed as this chapter was written.[13]

Since the subject matter of this research study does not include a thorough analysis of platforms and associations, and after giving a brief description of the issues experienced by this type of organisations, we will only refer to those including the brands under review because they are part of the sample selection criteria. These are, in alphabetical order, Asociación de Moda Sostenible de Barcelona, Asociación de Moda Sostenible de España (Catálogo), Asociación de Moda Sostenible de Zaragoza (Catálogo Deseclipsando), and Directorio Moda Impacto Positivo (Slow Fashion Next).

6 Five Brands: History, Raw Materials and Products, Supply Chain, Communication

When analysing the sustainable fashion landscape in Spain, the subject matter of this study, the following sample selection criteria were considered: payroll, turnover, membership in a sustainable fashion platform, territoriality, and readily accessible information. Table 2 shows the criteria followed on the basis of the Annex to the European Commission Recommendation dated 6 May 2003 (DOCE L 124 dated 20

---

[13]Statement posted on Facebook which evidences the disagreements between the different entities: In line with the unilateral decision of Asociación de Moda Sostenible de Andalucía (AMSA) to change its name to Asociación de Moda Sostenible de España (AMSE), Asociación de Moda Sostenible de Madrid and Asociación de Moda Sostenible de Barcelona are forced to issue a press release (see press release in Appendix III). Available on 5 April, 2017 on https://goo.gl/6CEmjd.

Table 2 Definition of the type of company based on balance sheet, annual turnover and number of employees

Annex to the European Commission recommendation

Type of company	Annual balance sheet (millions of euros)	Annual turnover (millions of euros)	Number of employees
Microenterprise	<2	<2	<10
Small SME	<10	<10	<50
Medium	<43	<50	<250
Large	≥43	≥50	≥250

Source Annex to the European Commission Recommendation, dated 6-5-2003 (DOCE L 124 dated 20-5-2003), in force since 1-1-2005. Prepared by the authors

May, 2003), to set parameters to business typologies. Online databases were also considered for all the other reference criteria.

Based on the reference data, the selected brands meet the microenterprise parameters. The sample brands have an annual balance sheet and turnover of less than 2 million euro, and less than 10 employees. Entrepreneurs do not have a background in the textile and fashion sector, but, based on the economic situation, they decided to make a career change in search of their own way, and jump to the sustainable fashion market. Table 3 shows a summarised description of the selected brands based on the sample selection criteria defined by the authors.

6.1 Brand 1

6.1.1 History and Spirit

This brand[14] embodies quite accurately the features of sustainable fashion microenterprises in Spain, which characteristics were discussed above and described in Table 2. The entrepreneur who created the brand does not have a background in the textile and fashion sector, but, based on the economic situation, she decided to make a career change in search of her own way, and jump to the sustainable fashion market.

In the case of Brand 1, both the insights of this conventional industry and the approach to sustainable fashion values helped have it on-board of this project. The brand selects fabrics, creates prints, and chooses colours and textures, as well as the style, sketch, design and pattern of their creations.

---

[14]The information about Brand 1 was taken from the brand's website and from a direct conversation with its leader.

Table 3 Summarised description of the selected brands based on the sample selection criteria defined by the authors

Name	Organisation date	Location	Products	Certifications	Type of company	Included in directories
Brand 1	2015	Spain	Unisex 0–3 years old baby clothes	GOTS, OOCGuarantee Oeko-tex Standard 100	Microenterprise	Directory: Moda Impacto Positivo, (Slow Fashion Next), Asociación Moda Sostenible España AMSE
Brand 2	2010	Spain	Bags	OCCGuarantee	Microenterprise	Directory: Moda Impacto Positivo, (Slow Fashion Next), Asociación Moda Sostenible España AMSE
Brand 3	2012	Spain	No gender women clothes	OCCGuarantee	Microenterprise	Moda Impacto Positivo (Slow Fashion Next) Moda Sostenible de Zaragoza (Deseclipsando)
Brand 4	2009	Spain	Unisex natural knits and dye clothes	GOTS and OCCGuarantee	Microenterprise	Moda Impacto Positivo (Slow Fashion Next)
Brand 5	2008	Spain	Women/men and children underwear	GOTS and OCCGuarantee	Microenterprise	Moda Impacto Positivo (Slow Fashion Next)
Brand 6	2015	Spain	Reused knit women and men clothes	GOTS and OCCGuarantee	Microenterprise	Directory: Moda Impacto Positivo (Slow Fashion Next), Asociación Moda Sostenible Barcelona

Source Prepared by the authors

6.1.2 Materials

Brand 1 uses OCCGUARANTEE[15] certified 100% organic cotton. It uses ecological digital printing with wash-resistant dyes free from toxic substances certified with Standard 100 by Oeko-Tex. This favours product recycling and reuse.

6.1.3 Business and Communication

The significance given by Brand 1 to recycling results in some business model innovations. For instance, customers can return their garments when they no longer wear them to bring new life to them. The brand is also planning to implement a rental system to buy a new or worn product and rent it in perfect condition so as to have any size available for a small monthly fee.

Brand 1 communications are posted on its website and social media, and they also appear in the media. The entrepreneur herself is in charge of this straightforward and original approach. The use of the first person to convey a clear message prevails in the communication speech.

A blog and a newsletter are some of the channels used to build customer relationships. Specifically, the website has buttons for customers to express their satisfaction with the products. In this connection, products are posted on the website based on the following elements: product name, price, customer assessments, use properties, healthy recommendations, and packaging characteristics—everything in terms of sustainability.

Moreover, the brand includes a statement that describes the sourcing and quality of raw materials, design processes—based on functional purpose and comfort—and unisex clothes, as well as its commitment to ecology and sustainability, particularly evidenced in the packaging which, in this case, stands out for its exclusiveness and customisation. It is worth pointing out the traceability of product manufacturing process data provided by the brand, which identifies all its suppliers and explains the characteristics of the products used.

6.2 Brand 2

6.2.1 History and Spirit

The designer of Brand 2[16] had been engaged in graphic design for many years, which stimulated a creative, methodical, and accurate mind. His love for nature helped shape an ecological soul, blending painting and fashion, which resulted in an environmen-

---

[15]Written in capital letters as used by the brand in its communications.

[16]All the information about Brand 2 was taken from the brand's website and from a direct conversation with its leader.

tally friendly brand. Through his collections of timeless bags, the designer offers everybody a personal touch which does not follow any trend. It is characterised by slow design in limited editions and with ecological fabrics. The brand also designs the prints of each collection as its main focus. It offers limited editions for customers running away from conventional designs, who are independent, environmentally friendly, and want their bag or rucksack to convey such spirit.

6.2.2 Materials

It works with OCCGUARANTEE certified organic cotton. Its processes are geared to generating less waste as per the Zero Waste philosophy. The leftover fabric is used to make products which require less material.

6.2.3 Business and Communication

From a business perspective, brand shaping is always complex, above all, when it comes to fitting in the role of players. In the case of Brand 2, products are manufactured by women from social inclusion programmes to promote social entrepreneurship and women empowerment, and they are sold on its online channel.

The brand's website and social media are the main communication channels. Its website, social media and newsletter are the tools used to convey its communication speech. In this case, it is characterised by a slow design message that promotes economy of proximity.

Just like the previous case, the project of Brand 2 began as a result of a personal entrepreneurial effort, which is evidenced and expressed in its communications.

Once again, product description refers to the use of less water resources and no pesticides during raw material production, among other characteristics for which it claims to be sustainable.

6.3 Brand 3

6.3.1 History and Spirit

Brand 3[17] was conceived as a sustainable fashion company that blends passion for fashion with the entrepreneurs and owners' life philosophy. It defines its clothes as ecological, sustainable and locally-made in decent working conditions, which makes entrepreneurs very proud. Most of its processes include hand manufacture. It works with timeless, easy-to-match patterns.

---

[17]All the information about Brand 3 was taken from the brand's website and from a direct conversation with its leaders.

Once we realised about the impact on nature, workers, and on ourselves, we decided to look for an alternative aligned with our way of living and feeling where respect for the environment, respect for workers' rights, and respect for the women who wear our clothes are essential.

6.3.2   Materials

Once again, the materials used are fabrics classified as ecological, sustainable and free from chemicals, which are purchased from GOTS-certified suppliers. To manufacture its clothes, Brand 3 mainly uses organic cotton, hemp and recycled plastic. The organic cotton used is grown in fields free from synthetic chemical pesticides, herbicides and fertilizers, and no toxic chemicals are used during spinning and production. Thanks to its multiple properties, hemp is a highly ecological raw material: it regulates body temperature; offers protection against UV and UVB radiation, thus alleviating the effects of the sun on the skin; and it has antibacterial, antifungal and antistatic properties. Moreover, it needs little water to grow. Since the plant is very resistant to plagues, it does not require pesticides or herbicides—or fertilisers, as it grows very fast.

6.3.3   Business and Communication

Brand 3 has a shop and a workshop at the same premises, and uses both the physical and the online channels to sell products. Every garment is hand-crafted, piece by piece, so that each item becomes unique and special. They produce a very small number of units per model—between 2 and 8—to imbue clothes with a strong identity load for the buyer.

The brand's communication speech is dominated by well-done work in which time and devotion play a leading role. Informative contents about the sustainable characteristics of their products prevail. For such purpose, the brand provides detailed information about garment care. It is very active on Facebook, Twitter, and Instagram.

It leverages product description as a vehicle to convey a sustainable message, highlighting the organic nature of raw materials, and the presence of certifications as sustainability assurance.

6.4 Brand 4

6.4.1 History and Spirit

Brand 4[18] is made up of a couple of designers who oversee design, dyeing, and cutting of their production manufacturing process. The female designer has a specific training in fashion, while the male designer contributes with his training in ecology, from green energies to waste recovery and reuse. They continue with their training in botany and organic chemistry to supplement their studies of handcraft natural dyes. Brand 4 is committed to sustainability and the slow fashion philosophy. It proposes a different viewpoint for the concept of fashion with ecological and healthy designs. It works with the conviction that the best outfit is the one which—in addition to style—has a minimum environment footprint.

They manufacture unisex items, made in organic cotton with the plant natural shades, that fit both the male and feminine figures. The brand also has a selection of basics. Basics are those items with a simple, comfortable, and timeless cut that we wear more often and make up the essentials of our wardrobe.

6.4.2 Materials

Creations are manufactured locally, using organic, hypoallergenic, biodegradable and compostable materials coming from environmentally- and worker-friendly crops with total traceability. Handcrafted prints are the brand's signature. They are made using plant- and flower-based natural dyes—each of them, different and exclusive—combined with natural shades of undyed cotton, such as beige, brown, and green. Fabrics and yarns are certified with the OCCGUARANTEE and GOTS stamps. The brand products are free from toxic substances, chemicals, chlorine, and synthetic dyes.

6.4.3 Business Model and Communication

Customers can purchase products at two physical points of sale and online. The brand produces by size and on demand (tailor-made). In this way, creations are customised to cater for all needs. This limited production involves great garment care both in every phase of production and in its finishes.

The brand offers a personal treatment. The "Create your perfect dress" project offers a choice between different alternatives to create customised items.

The brand makes an effort to ensure smooth communications with customers at all times, mainly through social media, newsletters, and video-call programmes. Based on the information posted on its website, and by combining different alternatives

---

[18]The information about Brand 4 was taken directly from the brand's website and discussed with its leaders.

customers can choose from various models "to their consumer preferences," thus tailoring the garment to the person instead of the other way round.

Product description includes material name, description, and characteristics; here, organic cotton as a certified raw material is given as assurance and evident sign of brand sustainability.

6.5   Brand 5

6.5.1   History and Spirit

Brand 5 (see characteristics in Table 2) is a small underwear clothing brand that works with sustainable and natural materials to offer alternatives to disposable fast fashion. With a track record in different arts and crafts, the entrepreneur of Brand 5[19] wanted to capitalise on her expertise by developing her own brand.

Her proposal is based on unique, comfortable underwear with environmental commitment. She seeks to create local, ecological, lasting and biodegradable consumption that takes into account every phase of production with zero waste.

The purpose of the brand is to take care of the fabrics that come in contact with the skin to ensure that they are free from toxic substances which might be absorbed.

6.5.2   Materials

The entrepreneur explains that she takes great care to select top quality, environmentally-friendly materials, and offers decent working conditions for the people involved in every process. The brand is committed to the highest certifications (GOTS) in organic materials, which ensure fibres from ecologically grown agricultural products.

Brand 5 works with Turkish cotton distributed in Europe: 95% organic cotton with 5% of elastane (Lycra), and 100% organic cotton for children collections. Moreover, light wool from Merino sheep is used in women underwear, while organic cotton, hemp and a waterproof, breathable synthetic fabric are used in cloth sanitary towels and pads.

Therefore,

> We try to get the GOTS certification for all our fabrics and, in those cases where we fail to get it, we choose suppliers which prioritise European production and transparency in every process. Most of our dyes have GOTS or OEKO-TEX certifications (we use non-aggressive dyes, free from heavy metals and chlorine-based bleach).[20]

---

[19]The information about Brand 5 was taken from the brand's website and from a direct conversation with its leader.

[20]Information available on Brand 5 website.

6.5.3 Business and Communication

Brand 5 uses its website both to place its products on the market and as a customer communication channel. Moreover, it works with a network of physical stores that share the brand's philosophy. There are 4 online points of sale and 19 physical points of sale—16 in and 3 outside Spain—with these characteristics.

Among the online communication tools, the brand uses the major social media (Facebook, Twitter, Instagram, Pinterest, You Tube and Vimeo) and its own webpage to for brand communication.

It creates plenty of text and image contents on its blog, as well as brand-produced videos to present products and relate them to the customer using stories.

The brand was also featured in national and international, mass circulation digital and chapter magazine articles, referenced to on its website.

Among the offline tools, the brand participates in fashion shows and conferences in which it promotes responsible consumption and zero waste generation. These conferences offer an excellent opportunity for the brand to convey knowledge and make it clear that it is targeted to a smart, reflective audience committed to responsible consumption.

It also provides for a space to learn about user/customer feedback and interact with them, making the written fabric certifications available to them.

6.6 Brand 6

6.6.1 History and Spirit

Brand 6 is an ecological and ethical fashion company founded by mid 2015 in Barcelona. It is led by a self-taught entrepreneur who is only interested in fashion for its environmental impact. The spirit of the brand was inspired by the upcycling practice, which is based on the motto "To turn waste into resources". In this way, sustainability criteria are part of her fashion designs.

This entrepreneur has a university education, with a strong international imprint from her previous professional experiences. While she worked in different sectors of this industry, she had always wanted to study fashion but could not do it. Year 2012 marked a turning point in terms of her opinion about fashion. At that time, she started reading and researching to learn about sustainable fashion and other alternatives with better environmental, social, and economic impacts. Since then, she has been engaged in the manufacture of garments for men and women, as well as accessories, using ecological materials—mainly recycled coffee sacks.

6.6.2 Materials

We use ecological and environmentally-friendly materials and processes. All the garments are hand-made, based on handcrafted cutting and tailoring processes. Proximity: since we are committed to the development of the local industry, we work with local suppliers, such as Organic Colours Cotton, to make sure that we use certified raw materials (see Table 2).

6.6.3 Business and Communication

The brand employs a handcrafted production process, using organic and recycled materials with minimum environmental footprint to give a unique, distinctive style to every item. While the online channel is used to place products on the market, there is a showroom at the workshop itself.

Clothes are designed along with other designers, creative experts, and artists to create unique pieces, raising awareness through cooperation with charities that promote social responsibility. The brand also intends to encourage fair trade, which ensures fair and decent working conditions that respect human rights. Another important aspect is the transparency used to make each part of the garment production process visible.[21]

The brand had a broad impact both on print media and the online environment—national and international trade and general-interest magazines. This was also possible because it is part of one the most active sustainable fashion platforms. In addition to those communication tools, the participation in sustainable fashion shows and events has played a major role, as it helped reach a high visibility level. The brand also uses social media and, mostly, its website to open dialogue with customers.

7 Interview Analysis

As shown in Appendix, interview responses were divided into five blocks. Against the model developed by Gardetti [2], the following criteria was set:

(a) Responses differentiate the aspects conceived by the brand under review from those "embraced" by the brand from other product value chain players, such as suppliers of raw materials, distribution/transport and marketing, etc.

(b) To claim that the brand is compliant with sustainability aspects, it is not enough to state that "the brand considered certain sustainability indicators": indicators need to be identified and followed by an explanation about how they are

---

[21]The information about Brand 6 was taken from the brand's website and from a direct conversation with its leader.

measured. These cases consider the variable "the brand takes into account sustainability aspects, but it does not measure them" based on the previous criterion; the brand claims that it takes into account sustainability aspects, but it does not measure them because it does not consider the indicators. For example, as to the efficient use of water, it is not enough to claim that a brand uses raw materials produced with efficient use of water. The brand should clearly specify how it does so, regardless of the practices—whether certified or not—of the other players involved in product traceability.

In connection with the response layout, in some cases brands answered the entire block of questions, or grouped the responses either because they did not had all the data or because they chose to bypass responses for confidentiality reasons. Brand 2 offers an example of this, as stated by its leader,

> As my brand is very small, you won't find useful data here. I implement sustainable processes, but I don't measure them. I implement plans based on growth. I am committed to looking for sustainable suppliers and products, but I don't gather the detailed statistics which -I believe-you need. Besides, I believe that this is internal data that I don't usually disclose.[22]

Later, Brand 2 answered that it believes that "too much information" is requested. While such a rich and unique response called for an analysis outside the framework provided for most of the interviews, we will refer to this brand as required.

Stakeholders[23]

Stakeholders play a major role in brand operation, just like brand awareness of who they are and their impact. When asked about their knowledge of their stakeholders and their ability to identify them—as shown in Table 4—only three (Brand 3, Brand 4 and Brand 6) out of the six brands under review were able to identify them in general terms. For the rest of them (Brand 1, Brand 2 and Brand 5), the stakeholder concept is not clear, and for this reason they can only identify their customers. In the case of the brands mentioned in the previous paragraph, this implies that they embrace other players' sustainable practices as their own. As an example, all the brands unanimously identified the efficient use of water by raw material producers, or the urge to respect social rights as if their own just because they are part of the value chain of brand manufactured products.

Environment

As shown in Table 4, this section describes the ways in which brands comply or fail to comply with environmental sustainability aspects, based on the criteria explained at the beginning of this section in items (a) and (b).

---

[22]Interview with the leader of Brand 2. This brand is included in the directory Moda Impacto Positivo of Slow Fashion Next, and in the directory of Asociación Moda Sostenible de España (AMSE).

[23]Stakeholders are understood as every individual or group which may affect or become affected by company goals.

Table 4 Brand stakeholders

Brands and stakeholders		
Brands	Are they able to identify their stakeholders?	List them
Brand 1	No	–
Brand 2	No	–
Brand 3	Yes	Organic material suppliers/producers, management/workers, sustainable product clientele/investors, local public administration
Brand 4	Yes	Suppliers, workers, customers
Brand 5	No	–
Brand 6	Yes	Customers, coffee importers, education centres, public administration

Source Prepared by the authors

The analysed brands—except Brand 2[24]—report a very different use of water. Brand 1, Brand 4 and Brand 5 rely on suppliers' certifications as assurance of sustainable use in raw material production and transformation, "This certificate currently ensures the most ecological means which uses the lowest volume of processing water".[25] As the brand passes on compliance to the other players, it is not compliant with this aspect. In addition to raw material certifications, Brand 3 claims that, "We collect rainwater, and we stock up and reuse most processing water".[26] Water plays a major role in the washing processes of recycled raw materials used by Brand 6. However, they do not take into account the water use impact. The brand explains that "as production volume grows, we will use industrial washing services to increase the number of sacks per wash, using shorter washing cycles".[27] They currently use regular washing processes.

The same happened with the efficient use of energy. Except for Brand 2, which has already stated that it does not measure sustainability aspects and, therefore, it cannot talk about them, all the other brands express their concern with the efficient use of energy, but they rely on certifications because they only associate it with raw materials. An example of this is the statement made by the representative of Brand 1,

---

[24]As explained at the beginning of the analysis, Brand 2 makes no claim about its sustainability assessment, though considers itself to be sustainable; hence the importance attributed by the authors to brand claims.

[25]Interview with the leader of Brand 1.

[26]Interview with the leader of Brand 3.

[27]Interview with the leader of Brand 6.

Just like in the case of water, the OCCGUARANTEE stamp also ensures the efficient use of energy in raw material growing, spinning, and manufacture, as the entire process is transparent and monitored.[28]

Only Brands 5 and 6 have implemented methods to control the use of energy, "We use energy-efficient light bulbs and, if possible, we reduce pressing and heating" (Brand 5). Brand 6 states,

A larger number of sacks per wash not only reduces the volume of water used, but it also reduces the number of washes and, therefore, energy consumption. We also consider the power companies we work with. We are currently working with a local company with green energy certification.

When asked about their waste recycling approach to reduce CO_2 emissions, Brand 1 explained that it implemented waste management and reuses fabric remnants and leftovers "to make the most of them," but it does not specify any amount. Brand 3 states that most of its work takes place in its own workshop to avoid transport to third parties, but it does not have it properly measured.

Moreover, Brand 4 sorts workshop waste and "discards hardly anything: leftover fabric is used to make smaller garments and, if they are not useful, they are taken to collection points". Brand 5 says that "we use energy-efficient light bulbs and, if possible, we reduce pressing and heating".[29]

As evidenced in the results of the interviews, there are multiple waste disposal systems in place. For example, Brand 6 explains that "we use both organic materials and leftover stock, in addition to recycled fabrics and combine them with coffee sacks. We also save the trimmings to make accessories or to add them to new designs".[30]

The use of toxic substances and materials in product manufacture is another aspect with a huge environmental impact. Once again, brands rely on certifications, but this case is different because the certification is directly granted to the brand. For example, Brand 1 explains,

In the printing process, we use digital textile printing, which is the most ecological alternative available. Our printing process is certified with Standard 100 by OEKO TEX. We are looking for even more ecological dyes in the near future.[31]

Brand 3 answered,

All our fabrics are ecological and have GOTS certification. No toxic substances are used in our workshop. Today, over 90% of our garments have no zipper, button, or any kind of accessory, but ecological fabrics.[32]

Brand 5 stated that, "We minimise the use of synthetic materials. Our goal is to eliminate the use of synthetic materials; we are analysing the use of rubber elastics".[33]

---

[28]Interview with the leader of Brand 1.

[29]Interview with the leader of Brand 3.

[30]Interview with the leader of Brand 6.

[31]Interview with the leader of Brand 1.

[32]Interview with the leader of Brand 3.

[33]Interview with the leader of Brand 5.

Faced with the question about their actions to eliminate the use of fossil fuels, brands answers were quite general. Brand 1 explains that it does not know because it has never considered that issue. Brand 3 claims to be "a local trade brand and, in a way, a KM0 brand".[34] Moreover, Brand 5 and Brand 6 refer to clean means of transport for product distribution or to the use of local suppliers to avoid polluting transportation for raw material delivery.

In connection with the use of Zero Waste technology, only Brands 1 and 5 state that they take it into account in the production process. As mentioned above, Brand 2 makes no reference to it, Brand 3 does not take it into account, and Brand 4 ignores the concept. Brand 6 will use it in the future in line with its plans to implement new technology.

Once again, brands rely on certifications regarding suppliers' use of soil, with labels and the brand's website as the ultimate channels to communicate the environmental impact of consumers' purchase decisions.

Brand 1 and Brand 3 offer alternative uses for the product when the consumer no longer needs it. In the first case, the brand leader explains,

> Our product labels encourage customers to return the garments when they no longer wear them, explaining that we will pick them up. The purpose of this initiative is to start a new second-hand sale and/or rental cycle at much more affordable prices.[35]

In the second case, "We offer the option to change some aspects of the garment free of charge and with no time limit, to give a fresh look and second use to the item".[36] Brand 4 makes no offering, while Brands 5 and 6 explain that they are planning to offer an alternative to close product lifecycle, but it is still in the making.

In connection with the type of waste management for leftovers, Brand 1 claims that "as mentioned above, we use all the leftovers to make other products, packaging, etc.".[37] In this connection, Brand 3 states that,

> We have virtually no leftovers because we don't work with collections; we could say that we work on a daily basis, with a very small number of garments. However, we alter some of the few garments which are left to make new items.[38]

Brand 4 also claims that it does not have much leftover production. Brand 5 states that small remnants are reused in smaller-sized items, and the rest is taken to textile collection points for reuse. In the case of Brand 6, as it produces,

> Many unique garments and very small editions of certain models to check market acceptance, we keep minimum stock. We try to avoid leftovers by producing on demand in order to minimise the use of unnecessary resources, materials, and services.[39]

Design and innovation

---

[34] Interview with the leader of Brand 3.

[35] Interview with the leader of Brand 1.

[36] Interview with the leader of Brand 3.

[37] Interview with the leader of Brand 1.

[38] Interview with the leader of Brand 3.

[39] Interview with the leader of Brand 6.

Based on the answers given by brand leaders, it is evident that—except for Brand 6—brands under review do not take into account recycling, reuse, or redesign in product design. Brand 5 is addressing this issue from the materials side. In the case of Brand 3 and Brand 4, they do not seem to have embraced design as an aspect to take into account for product sustainability. For Brand 6, the raw materials,

> jute and cocoa coffee sacks- are original. Once the sacks have fulfilled their carrying purpose, we give them an entirely different life with a new purpose. We implement a material "upcycling" process. This material turns from "waste" into one of our most valuable raw materials. These sacks cease to carry coffee and cocoa to dress our bodies and carry our belongings as accessories (bags, cases, etc.). In addition to coffee sacks, we use other garments, which we take apart and reuse for new designs.[40]

In essence, these brands have an underlying sustainability purpose, which they try to convey to their customers mainly through their products. This is how Brand 1 explains it: "It is part of the philosophy expressed through our products. Timeless, unisex quality products that last longer, have many more lives, and grow hand in hand with your baby before they get too small to fit". Brand 5 has also an interesting experience, "First, we worked more with this issue, but we realised that the message failed to take hold; now, we have decided to enhance the product so that it embodies our message, building trust in sustainable production". In the case of Brand 6, it explains that,

> Not only is our purpose to create clothes and accessories, but also to deliver seminars to discuss issues related to a sustainable lifestyle and responsible consumption. We have recently delivered our first seminar in our workshop called 'Reconnect with your essence; 21 stylists' tips to create a more sustainable wardrobe'.[41]

Another aspect is how to take into account the reappraisal of native culture and the consideration of other cultures in the design. In this connection, the brands under review express it differently. Brand 3 openly states that its design "doesn't follow that line".[42] While Brands 1, 5 and 6 have not consolidated it yet, taking into account that local culture is part of their identity.[43] The answers from the other brands about this aspect are quite general.

Innovation as an incremental improvement is the essence of disruptive design. This aspect is quite evident in Brand 1, which designs are focused on creating products that "last longer at times when they are usually short-lived because babies grow fast, making them as practical and convenient as possible".[44] Moreover, Brand 5 states that,

> We strive to improve item designs and to avoid underwires, frames and synthetic padding materials, both celebrating and leveraging the different types of body. We combine different materials to get a good finish, creating a comfortable, easy-to-wear, underwire-free bra.[45]

---

[40]Ibid.

[41]Ibid.

[42]Interview with the leader of Brand 3.

[43]Brands refer to the local culture as local urban culture.

[44]Interview with the leader of Brand 1.

[45]Interview with the leader of Brand 5.

The other brands do not have any disruptive strategy either for design or for production processes. This aspect is related to the investments or partnerships that brands managed to get in order to gain new skills or experiment with them. No brand complies with this aspect. This is aligned with the lack of serious consideration of sustainable awareness and practices, which results in a "shallow" use of the sustainable concept. Therefore, the term "sustainable" is only used as a way to improve the commercial strategy.

Social aspects

As shown in Table 5, in this section the brands provided very little, unspecific information. They are small companies where owners/entrepreneurs also play the worker/designer role. In the words of the leader of Brand 1: "So far, we haven't hired anybody".[46]

Only Brand 4 hires workers from time to time. And, with respect to suppliers, brands—once again—rely on certifications.

When asked about the Sustainable Development Goals, except for Brand 6, none of the brands knows or can identify these goals or the Global Compact.[47] An example of this is the answer given by Brand 1: "I honestly don't know what the compact is all about".[48]

In line with the above, the brands under review do not use external workshops for their production. As explained by Brand 6: "I haven't worked with external workshops yet, but I have visited some of them to build relationships in the future. Most of these workshops are part of social and labour reintegration programmes".[49]

All the brands willingly agree to comply with the standards, as per the different regulations. Since entrepreneurs are the only employees, corruption is only possible in terms of tax regulation compliance. As to suppliers, they leave everything up to certifications. Brand 1 claims that "we only work with companies that we fully trust, such as Organic Cotton Colours, trusted cooperatives, etc."[50]

They are all aware of their responsibility for showing the impact of fast fashion. In this connection, Brand 5 states that,

> As I explained before, we first had worked more with this issue, but when we realised that the message failed to take hold, we focused on improving and making top quality garments, so that customers trust us and pass on our data. I have been in this industry for a long time and I can see that the message doesn't take hold—giving lectures on sustainability is very different from having a sustainable company. We focused on creating a sustainable company to promote sustainability, but we are tired of the gurus.[51]

Consequently, we identify different ways of action: Brand 3 explains its *modus operandi* as follows, "We try to explain the positive aspects of sustainable fashion

---

[46]Interview with the leader of Brand 1.

[47]Global Compact, also known as UN Global Compact.

[48]Interview with the leader of Brand 1.

[49]Interview with the leader of Brand 6.

[50]Interview with the leader of Brand 1.

[51]Interview with the leader of Brand 5.

Table 5 Brands and their relationship with all the aspects inherent to sustainability

Brands	Water	Energy	Soil	CO$_2$	0% Waste	Hazardous materials	Information consumers	Responsibility fast fashion	Stock/donation
Brand 1	–	–	–	/	+	+	+	+	+
Brand 2	■	■	■	■	■	■	■	■	■
Brand 3	–	–	–	/	–	+	+	+	+
Brand 4	+	–	–	/	–	+	+	+	+
Brand 5	–	+	–	/	/	+	+	+	+
Brand 6	–	+	–	/	–	■	+	+	+

Brand	Social aspects						
	Human rights	Labour rights	Human trade	Modelling	Inclusion	Anticorruption	Fast fashion?
Brand 1	/	/	/	+	–	/	+
Brand 2	■	■	■	■	+	■	■
Brand 3	/	/	/	+	–	/	+
Brand 4	/	/	/	+	–	/	+
Brand 5	/	/	/	+	–	/	+
Brand 6	/	/	/	+	–	/	+

(continued)

Table 5 (continued)

Brand	Marketing			Raw materials		
	Responsibility	Communication	Customer relation	Raw materials	Animals	Recycling
Brand 1	+	+	+	Organic cotton	■	+
Brand 2	■	■	■	Organic cotton	■	■
Brand 3	+	+	+	Organic cotton, organic hemp and recycled plastic	■	+
Brand 4	+	+	+	Organic cotton, natural dyes, jute, wool, hemp, style	■	+
Brand 5	+	+	+	Organic cotton, merino wool, elastane	■	+
Brand 6	+	+	+	Jute, organic cotton	■	+

Note + the check-marked complies with the aspect

– the check-marked does not comply with the aspect

■ the check-marked aspect is not mentioned by the brand

/ the check-marked aspect is mentioned, but they do not measure it

Source prepared by the authors

against fast fashion directly at the store".[52] Brand 4 explains that it is the regional coordinator of the Fashion Revolution Day movement.[53] Brand 6 states that the seminars that it is planning to hold "will promote a positive way to make fashion; and all our communications on social media should be positive, along the same line".[54]

None of the brands declared to make any kind of social inclusion in the product design, creation, or distribution processes. In this connection, Brand 1 states that its principles are

> To strive for a decent and encouraging job throughout the chain; to have the utmost respect for the environment so that tomorrow our kids will have a place to live, breathe, and enjoy; to protect babies' health; and, ultimately, to design products that make both their and our lives better.[55]

In general, we can say that all of them, in the words of Brand 2, gradually "implement plans based on growth".[56]

Communication and marketing

The business model of the brands under review has a strong digital basis, as evidenced in the answer of Brand 5: "Since we can't have a physical store, we decided that our website was the store that we needed, and made ourselves known in fairs and events so that word-of-mouth recommendations lead customers to our website",[57] they say. When asked about the rationale of their websites and the role that they play, they unanimously answered: to communicate and sell our brand and products (Brand 1, 2, 3 and 4), in addition to be "a distribution channel for our products," as stated by Brand 6.[58] Therefore, their websites play an essential role in their communications.

Regardless of the number of followers, brand activity on social media and update frequency are really important, as shown in Table 6.

Brands have a clear stance on using social media as communication tools. Moreover, Brand 5 considers that they "have become the shop window where we say that we're still alive and working".[59] Brand 5 recognises that updates are interesting and, particularly regarding newsletters, it believes that they require "great effort".[60]

As to content priorities, they recognise the high commercial profile of the environmental and social information.

An interesting aspect of the offered contents are the issues related to product use and care. These issues are generally addressed on the brands' websites. For example,

---

[52] Interview with the leader of Brand 3.
[53] Interview with the leader of Brand 4.
[54] Interview with the leader of Brand 6.
[55] Interview with the leader of Brand 1.
[56] Interview with the leader of Brand 2.
[57] Interview with the leader of Brand 5.
[58] Interview with the leader of Brand 6.
[59] Interview with the leader of Brand 5.
[60] Ibid.

Table 6 Communication channels used by brands and update frequency

Brand	Communication channel	Update frequency
Brand 1	Social media	Daily
	Newsletter	Monthly
Brand 2	Social media	No answer
	Newsletter	No answer
Brand 3	Social media	Daily, on business days
	Newsletter	None
Brand 4	Social media	Weekly
	Newsletter	Once every four months
Brand 5	Social media	Weekly
	Newsletter	None
Brand 6	Social media	Twice a week
	Newsletter	Monthly

Source Prepared by the authors

Brand 1 explains that "the technical information about care, among others, is available both on our website, under product categories, and on the label".[61] However, Brand 3 provides this information verbally at the store: when a new customer comes in, "we talk to them, offering detailed information about the best way to wear and take care of the product".[62] Brands 4 and 5 point out that they explain how to take care of the garments both directly in conversations with the customers and on their websites. However, Brand 6 has not posted such information on its website, but it is planning to do so, "Yes, we will add product care information on our website, in the online store. Our webpage will have a product wear section with customers' style ideas and suggested looks".[63]

In this connection, Brand 1 states that "everything is part of how we convey our message and our sustainable philosophy. We use the same channels that we explained above". In turn, Brand 3 does not answer the question. Brand 6 is aware of its lack of information, but it categorically replies that it "doesn't provide information".[64] Brands communicate their leftover and remnant management on their websites. In this case, Brand 1 believes that its information efforts are not enough, as it argues that such information is "part of the organic cotton we use in all our products".[65] Brands 2 and 3 do not answer the question, while Brands 5 and 6 recognise that they

---

[61] Interview with the leader of Brand 1.

[62] Interview with the leader of Brand 3.

[63] Interview with the leader of Brand 6.

[64] Interview with the leader of Brand 3.

[65] Interview with the leader of Brand 1.

do not take any action in this connection. In turn, Brand 4 says that "customers can witness leftover and remnant management mechanisms in situ".[66]

The same happens with supply chain information. Brand 1 claims that it provides all the information available in the environmental section of its website, under the "About us" tab. Brand 3 provides information about its only supplier, and the rest is on display at the store. Brand 4 uses labels as its communication channel with customers to inform about product traceability. The web is the communication channel used by Brand 5 to inform about "the used processes and materials, the chosen processes and materials". And, finally, while Brand 6 provides some information on its website, it chooses direct information at its workshop as the communication channel for traceability information.

In all cases, websites and direct information are—to a certain extent—the shop window to communicate product traceability data.

Raw materials

Raw materials is one of the issues most referenced to by brands, and the one sustainability most relates to. As to raw materials, brands were asked if they take into account the circular economy criteria in product design, and to explain how. Responses are varied. On the one hand, Brand 1 asserts that it takes into account these criteria, as evidenced in their pro-recycling garment return initiative and second-hand sale and/or rental cycle. On the other, Brand 3 turns to compostable organic products manufacture and waste management (recycling and reuse). Brand 4 takes these criteria into account, buy recognises that it has not implemented them yet. And, finally, Brand 6, which uses waste as raw materials, is analysing how to close the loop.

As to prioritising local raw materials to design their products, their scenario is diverse. In some cases, brands argue that they do not take that aspect into account (Brand 1), or do not answer the question (Brand 2, Brand 3). Brands 4 and 6 claim that—and describe how—they take it into account. Brand 4 answers, "Yes: salt, carob, and different flowers for dyes and eco-printing, etc.". However, in the case of Brand 6, this aspect is more related to raw material suppliers than to the raw material itself, as they claim that "we prioritise local raw materials, like organic cotton and organic hemp from local or domestic companies, such as Organic Cotton Colours".[67]

Brand 5 gave a very interesting answer, showing willingness to implement this practice, but with current limitations forcing it to take a different path.

> Yes, we take it into account, but it isn't viable at this moment because in Spain there aren't Spanish organic fabrics; elastics, zippers and metal accessories are national and manufactured in the country as per the OEKO-TEX certification. Our fabrics are European -Turkish- and hold GOTS certification.[68]

---

[66]Interview with the leader of Brand 4.

[67]Therefore, the transformation process of raw materials—cotton traded by Organic Colours Cotton—is certified by GOTS. Driven by their desire to provide sustainability assurances for raw material production, they offer their own certification. This is possible because their business model traces product back to sourcing through to distribution.

[68]Interview with the leader of Brand 5.

The answers to questions related to use of trimmings, vintage elements, used materials or samples, as well as their use frequency and whether or not they are part of collections, turned out to be very insightful, since brands provided valuable information for this study. This was the case with the answer given by Brand 1: "In our case, we can't use recycled fabrics because we need fully toxic-free organic fabrics, and we can't find second-hand fabrics of this type".[69] Brand 3 gave no answer, and Brand 4 answered, "Yes, but not very often".[70]

Brand 5 claims that "we use large trimmings in smaller products—nursing pads, sanitary towels, etc.".[71] And, of course, recycling is the main raw material source for Brand 6: "Everything depends on the materials available when we start working. We try to reuse and make the most of our resources".[72]

The importance that brands attribute to certification is probably one of the key questions of this study. In line with their current discourse, answers range from Brand 1 arguing that it is important "to ensure material and product transparency, ethics, sustainability, ecology, purity, and quality," to Brand 3 regarding certification as "essential" or Brand 4 pointing out that it helps "build consumer trust".[73]

However, the most interesting answer was given by Brand 5, which agrees with the other brands in that they inspire trust, but adds a remark to take into account: it avoids "products coming from Asia as certifications are often altered".[74] This provides relevant information when it comes to assessing certification authenticity. Finally, the opinion of Brand 6 also involves a qualitative contribution, as its leaders assert that "certifications are good, but they can be very expensive",[75] an aspect that should be considered, as it restricts brand affordability.

Supply chain

For product sustainability assessment, it is essential to know the traceability of the supply chain and every process involved in product creation. In this connection, Brand 1 and Brand 4 only state that they take it into account. For Brand 3, to have its own workshop to carry out all its processes is virtually a proof of "100%" transparency.[76]

Brand 5 also claims that it takes traceability into account throughout the process as far as it is within its scope, "from the fabric arrival to the product delivery to the end user".[77] In this connection, Brand 6 is aligned with Brand 5, as it states that "we know the supply chain of certain materials from beginning to end, but some others

---

[69]Interview with the leader of Brand 1.

[70]Interview with the leader of Brand 4.

[71]Interview with the leader of Brand 5.

[72]Interview with the leader of Brand 6.

[73]Interview with the leader of Brand 4.

[74]Interview with the leader of Brand 5.

[75]Interview with the leader of Brand 6.

[76]Interview with the leader of Brand 3.

[77]Interview with the leader of Brand 5.

are recycled, which makes it difficult to learn about material traceability before we received them. The cutting and tailoring process is controlled in our workshop".[78]

When asked about their awareness of the environmental and social impact of the elements involved in product design, production and distribution, all the brands, except for Brand 5, answered "yes" to the question. Brand 5 explains that it "tries to be aware of everything, but some aspects are beyond our control; as regards textile production, we must rely on certifications".[79] And the leader of Brand 6 says,

> As a company, I believe in the implementation of the triple bottom line: economic, social, and environmental. We need to develop and implement the tools to assess the positive impact. For such purpose, we partnered with a business development platform that shares these values and goals. I know that everything begins with eco-design and the initial product conception considering its entire life cycle, followed by ethical production and different distribution-related aspects. We should be consistent and apply the same principles and values both to every dimension of business development and to the conception and development of the products and services that we offer; and to our personal daily lives as well.[80]

As explained above, the answers related to the verification of the transparency degree in the different phases of the process were varied. Two brands gave no answer (Brand 2 and Brand 3); one of them just asserted it (Brand 4); and Brand 1 and Brand 5 only identified the raw material production process as verified by certifications. In the words of Brand 1's representatives, they work "closely" with suppliers and they are "fully familiar with their project, plus they have all the necessary certifications".[81] Brand 5 relies on certification and it usually asks suppliers.

Moreover, Brand 6 states that the company's sole proprietorship ensures transparency, "As this is a microenterprise, there is transparency in every process".[82]

Annual income (in €)

To assess economic sustainability, brands were asked about profitability (net after tax, expressed as a percentage of income), and about their growth prospects for 2018 and 2019, as compared to 2017. They were also asked about their investments at business start-up and consolidation, and about the average price of their products. Brand 1 and Brand 2 did not provide such information as they did not consider it appropriate; "besides, I believe that this is internal data that I don't usually disclose".[83] The other brands agreed to our request on condition that the information was not disclosed along with their names. Therefore, we will address the data as generally as possible, and only to the extent of value addition to the subject matter of this study. All the brands are in the growing phase, and their average initial investment amounts to € 20,000, with € 87 as average price for their products.

We cannot conduct a more thorough analysis with the data provided.

---

[78] Interview with the leader of Brand 6.

[79] Interview with the leader of Brand 5.

[80] Ibid.

[81] Interview with the leader of Brand 1.

[82] Interview with the leader of Brand 6.

[83] Interview with the leader of Brand 2.

8   Conclusions

Is it enough to hold an organic cotton certification to claim that a brand is sustainable? This question was the starting point of our research study, and given the interest of brands in claiming themselves as sustainable, this turned to be quite worrisome.

After gathering free-access information provided by the brands on their websites and other media collected on them, and having interviewed brand leaders and compared results to the model developed by Gardetti [2], we may conclude that, while using certified organic cotton appears to be a major element for brands to claim themselves as sustainable, it is not the only aspect considered to define themselves as such.

Based on the analysis of the brands under review, we may conclude that microenterprises have a deeper knowledge and are more familiar with sustainability certifications than with sustainable aspects, measuring tools, and controls. However, they seem to understand that only large companies can afford to measure and control the sustainability level, unaware of the fact that they can do so with the models available to them, as they are designed to apply to any undertaking, regardless of number of employees or turnover. Another important aspect to point out is the trend to embrace the sustainable aspects from other players in the product supply chain, such as suppliers of raw materials, distribution/transport and marketing as their own. For example, communication and marketing contents developed by brands highlight the use of certified organic cotton under product description. Brands leverage product description as a vehicle to convey a sustainable message, underscoring the organic nature of raw materials and reporting about their ancillary material suppliers.

The answers show that, so far, brand commitment to the efficient use of water and energy is merely symbolic and only associated to raw materials, disregarding the direct implications on all the other processes until the delivery to the end user.

While brands are positively willing to comply with sustainability criteria, they lack information to properly identify them, evidencing a strong intuitive approach. The sector still lacks information, maturity and, thus, consistency between the aspects involved in sustainability and reality.

Brands take sustainable aspects partly into account, and fail to take any quantitative measures. CO_2 emissions are an example of this. Therefore, this ends up as a mere statement of good intent.

In connection with transparency and self-assessment ability, we may say that the sector is moving really slowly. In this respect, marketing criteria have taken a particularly deeper hold.

Associations and platforms' reference to certifying agencies and, ultimately, to the disclaimer regarding the accuracy of the information provided by brands as membership criteria evidences that the business side is more important than the fact that being a member of such agencies accounts for other purposes that are more important than sustainability-defining aspects. Moreover, to become part of these organisations seems to be prioritised over the sustainability aspects.

The answers given by the brands show the struggle against the search for transparency suggested by Gardetti [2]. Brands fail to provide (hide?) information. Brands provide specific information and have direct data to report only in the sections related to raw materials and marketing (with contents that they consider important for consumers). In the other sections, they do not have—or do not want to report—such data, or they turn to other players in the supply chain to embrace them as such. Therefore, any change of letting information flow as a sustainability tool fades away.

Transparency and control go hand in hand. This research study highlights the absence of control by external agencies, regardless of the certifications for certain sustainability aspects. In this field, sustainable fashion platforms and associations play a sort of certifying role in the eyes of the consumer. As they inspire credibility and encourage the consumption of member brands, can sustainability be achieved without holding any certification? Of course.

Finally, based on the analysed data, leaders lack information that would prove useful to reflect on sustainability aspects and, consequently, to set and measure indicators. Moreover, brands are not aware of the potential benefits of reflection as an opportunity to improve through assessment.

Sustainable entrepreneurs' lack of thorough knowledge of each of the aspects involved in sustainability calls for an important training initiative to align brands sustainable construct with the "Big Picture" of sustainability.

Looking Forward

Since these brands use raw materials which partial processing, in some cases, in Turkey, the authors suggest investigating the impact of refugee exploitation on both production and brands.

To assess the foundations of the criteria used by sustainable fashion organisations may also prove interesting.

Based on the results of this research study, we suggest analysing the reasons why consumers take the reports of these organisations as audited certifications, disregarding the proper external auditing processes. A sole proprietorship does not entail being honest with themselves, the public administration, the taxing system or other players in the value chain.

The authors of this research study agree with Lise Kingo, executive director, United Nations Global Compact, who closed the Global Compact meeting held in Buenos Aires on 25 and 26 April 2018 saying that the Sustainable Development Goals pose a huge global challenge short of time. The time of wishing is over, it is time for deeds.

Appendix

No.	Question
1	**Stakeholders**
1. 1	Do you know your stakeholders? Stakeholders are understood as every individual or group which may affect or become affected by the company goals. Can you name them?
2	**Environment**
2.1	Does your brand take into account the efficient use of water? How?
2.2	Does your brand take into account the efficient use of energy? How can this be checked?
2.3	Do you take into account waste recycling to reduce CO_2 emissions? How?
2.4	Do you avoid the use of toxic or hazardous substances and materials to manufacture your products? How? If not, what should you change?
2.5	What actions have you implemented to eliminate the use of fossil fuels?
2.6	Do you use the Zero Waste technology to manufacture or make your product? What kind of technology?
2.7	What kind of information do you have regarding your suppliers' use of soil?
2.8	Are consumers aware of their environmental responsibility in their purchase decision? How?
2.9	Do you offer alternative uses for the product when the consumer no longer needs it?
2.10	What type of waste management do you have in place for leftovers?
3	**Design**
3.1	Do you take into account recycling, reuse, or redesign in product design?
3.2	Do you try to communicate the sustainable culture through your brand products? How?
3.3	Do you take into account the reappraisal of native culture and the consideration of other cultures in the design? Which?
3.4	Do you use disruptive design in your products? Disruptive design is understood as innovation that leads to an incremental improvement. What innovations did you implement?
3.5	Did you invest or partner in order to gain and experience new competencies? Specify which?
4	**Social aspects**
4.1	Do you take into account the legislation when it comes to workers' compensation? How? Both for direct and indirect workers?
4.2	Do you take into account the legislation on health and safety in the workplace? Specify actions. Consider outsourced workers as well
4.3	Do you take into account the legislation on working hours? Specific aspects; consider outsourced workers as well

(continued)

(continued)

No.	Question
4.4	Do you take into account the Global Compact Sustainable Development Goals? Specify the goals that you prioritised
4.5	Do you have specific anti-corruption protocols in place in your company? Which?
4.6	Do you communicate with your stakeholders? What channel do you use?
4.7	Do you somehow acknowledge the fast fashion issue? How do you communicate it to your customers?
4.8	Do you make any kind of social inclusion in the product design, creation, or distribution processes? How?
5	**Communication and Marketing**
5.1	Does your brand have a website? What is its purpose?
5.2	Is your brand active on social media? What social media is it on? How often does it update them? How many followers does it have?
5.3	Does the brand use a Newsletter for customer communication? How often?
5.4	Does the brand prioritise information contents in its marketing tools? What aspects does it highlight?
5.5	Does the brand prioritise commercial contents in its marketing tools? What type of contents does it use?
5.6	Does the brand offer information about product use and care? Through what channels (website, newsletter, physical store, social media)? What kind of information?
5.7	Does the brand offer information about the impact of product purchase? If yes, specify
5.8	Does the brand offer information about leftover and remnant management? What channel does it use? How?
5.9	Does the brand communicate customers its efficient use of energy? What information does it offer?
5.10	Is supply chain information available to customers? How?
6	**Raw Materials**
6.1	Does the brand take into account the circular economy criteria in product design? Which?
6.2	Does the brand take into account local raw materials in product design? Can you name them?
6.3	Does the brand use fabric trimmings and old pieces to make a new garment? How often?
6.4	Does the brand use samples and used materials to make a new garment? Are these garments usually part of collections?
6.5	Are raw materials original or recycled? Are they certified?
6.6	How much importance do you attribute to certifications?
7.	**Supply chain**
7.1	Do you know the traceability of the supply chain and every process involved in product creation?

(continued)

(continued)

No.	Question
7.2	Is the brand aware of the environmental and social impact of the elements involved in product design, production and distribution?
7.3	Can you verify the transparency degree in the different phases of the process?
8.	**Annual income (in €)**
8.1	Profitability (net after tax, expressed as a percentage of income)
8.3	What are the brand's growth prospects for 2018 and 2019, as compared to 2017? (Expressed as a percentage against 2017)
8.4	What was the initial business investment? (in €)
8.5	What is the average price of your products? (in €)

References

1. Riera S (2016) Las claves de un cambio sistémico. In: Moda sostenible: La nueva hoja de ruta del sector. https://www.modaes.es/visor-online.php?id=58&name=Modaes.es+Dossier+-+Moda+sostenible#1. Accessed 12 Apr 2018
2. Gardetti MÁ (2017) Textiles y moda ¿Qué es ser sustentable? LID Editorial Empresarial, S.R.L., Buenos Aires
3. Comunicarse (2018) La ONU califica la industria de la moda como una "emergencia ambiental". http://www.comunicarseweb.com.ar/noticia/la-onu-califica-la-industria-de-la-moda-como-una-emergencia-ambiental. Accessed 9 Apr 2018
4. Algayerova O (2018) Fashion and the SDGs: what role for the UN? UNECE. https://www.unece.org/fileadmin/DAM/RCM_Website/RFSD_2018_Side_event_sustainable_fashion.pdf. Accessed 9 Apr 2018
5. Hernández R, Fernández C, Baptista P (2010) Fundamentos de metodología de la investigación. McGraw-Hill/Interamericana de España, S.A.U., Madrid
6. Riaño P (2016) Hacia un nuevo paradigma. In: Moda sostenible: La nueva hoja de ruta. https://www.modaes.es/visor-online.php?id=58&name=Modaes.es+Dossier+-+Moda+sostenible#1. Accessed 12 Apr 2018
7. Deloitte (2015) Sustainability and compliance trends. https://www2.deloitte.com/nz/en/pages/risk/articles/sustainability-and-compliance-trends.html. Accessed 27 Feb 2018
8. GLOBAL RESOURCE INITIATIVE (2015) Guía para la elaboración de memorias de sostenibilidad. https://www.globalreporting.org/Pages/resource-library.aspx?resSearchMode=resSearchModeText&resSearchText=G4&resCatText=Reporting+Framework&resLangText=Spanish. Accessed 1 Mar 2018
9. Fletcher K (2012) Gestionar la sostenibilidad en la moda. Blume, S.L., Barcelona
10. Club De Excelencia En Sostenibilidad, Nielsen (2009) Consumo Responsable y Desarrollo Sostenbile ¿qué opinan los españoles? http://www.clubsostenibilidad.org/publicaciones/consumo-responsable-y-desarrollo-sostenbile-que-opinan-los-espanoles/. Accessed 1 Mar 2018
11. Jiménez Herrero LM (2017) Desarrollo sostenible. Transición hacia la coevolución global. Ediciones Pirámide, Madrid
12. Alonso N (2016) TEXTIL "BIO" Guía de certificaciones para orientarse un poco. http://vidasana.org//noticias/textil-bio-guia-de-certificaciones-para-orientarse-un-poco. Accessed 27 Feb 2018
13. IED SOSTENIBILIDAD (2015) Etiquetas y Sellos Certificados. https://sostenibilidad.iedmadrid.com/bibliografia-y-documentacion-sostenible/certificaciones-y-normativas-sostenibilidad/etiquetas-y-sellos-certificados/. Accessed 27 Feb 2018

14. Aitex (2018) Certificado made in green by OEKO-TEX. http://www.aitex.es/certificado-mad e-in-green-by-oeko-tex/. Accessed 7 Apr 2018
15. Modaes.es (2016) Dossier Moda Sostenible: la nueva hoja de ruta del sector.2. https://www.mod aes.es/visor-online.php?id=58&name=Modaes.es+Dossier+-+Moda+sostenible#1. Accessed 12 Apr 2018
16. OEKO-TEX (2018) OEKO-TEX® I STANDARD 100 by OEKO-TEX®. https://www.oeko-t ex.com/en/business/certifications_and_services/ots_100/ots_100_start.xhtml. Accessed 7 Apr 2018
17. Christensen Clayton M (1997) The innovator's dilemma. When new technologies cause great firms to fail. Harvard Business School Press, Boston
18. Gjerdrum Pedersen ER, Reitan Andersen K (2013) The SocioLog.dx experience: a global expert study on sustainable fashion. Copenhagen Business School and Mistra Future Fashion, Copenhagen
19. Hart SL, Milstein M (1999) Global sustainability and the creative destruction of industries. MIT Sloan Manage Rev 41(1):23–33
20. Papanek V (1971) Design for the real world. Human, ecology and social change, 2nd edn. Chicago Review Press, Chicago
21. Teece DJ, Pisano G, Shuen A (1997) Dynamic capabilities and strategic management. Strateg Manage J 18(7):509–533
22. Williams D (2015) Fashion design and sustainability. In: Blackburn RS (ed) Sustainable apparel. Production, processing and recycling. The Textile Institute and Woodhead Publishing, pp 163–185

Naturally Colored Organic Cotton and Naturally Colored Cotton Fiber Production

Gizem Karakan Günaydin, Ozan Avinc, Sema Palamutcu, Arzu Yavas and Ali Serkan Soydan

Abstract Processing of fibers into textile materials requires the usage of extensive water, energy, chemical and other related resources. Dyeing processes may cause environmental pollution due to its chemical dyestuff and dyeing process auxiliary usage. There are some new considerable efforts for reducing the ecological hazard and waste generated during textile processing or developing sustainable and green materials. One of these promising approaches is to promote the usage of naturally colored cotton fiber usage and its production. As the world is moving towards to the pollution-free organic textiles and products, the naturally colored cotton fiber is going to be the next buzz word in the textile market. Since, the production process of naturally colored cotton skips the most polluting activity (dyeing) of the textile product manufacturing. Indeed, naturally colored cotton fiber usage for textile materials can eliminate the need for dyeing process due to their inherent color characteristics leading to water, chemical and energy savings with no synthetic dye usage for coloration. Not only the cultivation and the usage of the naturally colored cotton fiber but also the cultivation and the usage of naturally colored organic cotton fiber have also recently been increased. For example, brown and green naturally colored cotton fibers can be grown organically or conventionally. Naturally colored cotton growers have less requirement for the pesticides, insecticides since these varieties have already insect and disease-resistant, salt-tolerant qualities as well as they exhibit property for drought. Organic agriculture is a production management system which increases biodiversity as well as soil biological activity. This production is based on the applications of maintaining and enhancing the ecological harmony. Naturally colored organic cotton fiber (NACOC) has been a conspicuous textile fiber as the social trend of eco-friendly living has increased. NACOC fibers are naturally pigmented fibers for some limited color shades such as green, brown, mocha and red and their relevant shades. The color variety depends on the gene of the fiber as well as the seasons and geographical locations due to climate and soil variations. NACOC has

G. K. Günaydin
Buldan Vocational School, Pamukkale University, Buldan-Denizli, Turkey

O. Avinc · S. Palamutcu · A. Yavas · A. S. Soydan (✉)
Textiles Engineering Department, Pamukkale University, Denizli, Turkey
e-mail: assoydan@pau.edu.tr

© Springer Nature Singapore Pte Ltd. 2019
M. A. Gardetti and S. S. Muthu (eds.), *Organic Cotton*, Textile Science and Clothing Technology, https://doi.org/10.1007/978-981-10-8782-0_4

high resistance to insects and diseases. There have been some investigations for an improvement for the genetic properties of naturally colored cotton fibers in respect of better yield and better fiber qualities regarding strength, length and micronaire since natural colored cottons are desired to be more competitive against conventional white cottons. NACOC has a cost advantage with the elimination of dyeing process in fabric manufacturing. Additionally, instead of color fading problem which can be encountered in the case of dyed white cotton fibers, the color of the naturally colored cotton fiber becomes stronger after laundering. It has also been declared that clothes made from NACOC have been successful for preventing skin diseases as well as protecting skin from ultraviolet radiation. A significant number of research works have been carried out on white cottons, while naturally colored cottons were used to be left behind. Though, when environmental pollution has started to be one of the most urgent and important problem of the world, naturally colored cotton fibers are one of the more preferred options for more sustainable, renewable and ecological textile production. White cotton fiber is one of the most chemically intensive crops cultivated. Though grown on 3–5% of the world's farmland, it is liable for the usage of 25% of the world's pesticides. For these aforementioned reasons, organically grown naturally colored cotton fiber has attracted a massive attention over the last few years. In this chapter, an elaborative review of naturally colored organic cotton fibers, naturally colored cotton fiber types, their properties, their production and their recent developments from a broad perspective and with many different angles is given in detail.

Keywords Naturally colored cotton · Organic cotton · NACOC · Brown cotton Green cotton · Gossypium hirsutum

1 Introduction

Cotton as the main consumable natural fiber type has been cultivated in different areas of the world. Main cotton fiber growing countries in the world can be listed as India, China, United States, Pakistan, Brazil, Uzbekistan, Australia, Turkey, Turkmenistan and Burkina Faso. The major advantages of the cotton which make it prominent comprise wearing comfort, natural appearance, moisture absorbency as well as its renewable status. Its importance of economic role in many producing countries is also the main reason for why it has been so popular and widespread among the world for years. Furthermore, cotton farming provides employment in many sectors such as ginning, storing, utilization of farming equipment etc. However, some disadvantages of contamination introduced during harvesting, ginning and handling lead to quality and price variations among the different regions which require standardization. 90% of cotton agriculture is conducted in the in northern hemisphere and the land ratio is about the 2.5% of agricultural land of the world. Cotton's share of the world textile fiber market is about 37%. China (mainland), India, Pakistan and Turkey are considered to be remaining major textile economies with a dependency on cotton imports [1–3].

The term of sustainability has been used frequently in textile sector. Organic production is a way of providing the sustainability with balancing a growing economy, protection for the environment and social responsibility which lead to an improved quality of life for the future generations. Organic cotton is the main raw material for enhancing the textile sustainability. Other common terms used for organic cotton fiber, especially at the beginning of the production are green cotton, biological cotton and eco-friendly cotton. The conventional cotton consumes too much pesticide/crop protection products with many adverse affects on the environment. Moreover as the cancer risk has increased, the consumers are more rigorous about the raw materials of the end-products which do not have any chemical residues. Organic cotton cultivation is similar with conventional cotton cultivation however it differs in some aspects. Synthetically compounded conventional cotton production chemicals (fertilizers, insecticides, herbicides, growth regulators and defoliants) are forbidden to be used for at least three years before organic cotton production begins. Hand picking is mostly preferred as the same with conventional cotton harvesting. Although organic cotton cultivation has advantages in many aspects, there are some limitations in organic cotton production. When the variaties that are suitable for high fertilizer use are grown under organic conditions, there are high yield losses. This result discourages the farmers from continuing the organic cotton production. Additionally, organic cotton production is more challenging with the elimination of fertilizers, pesticides and other agrochemicals which require higher prices compared to conventional cotton. Although there are many countries claiming that they cultivate organic cotton, they can not be named as "organic cotton" because of inadequate number of certifying organizations. It is thought that the trend for organic cotton cultivation will increase sharply as the awareness of sustainability increases among the world. It should be remembered that although conventional cotton is grown on 3–5% of the world's farmland, it is liable for the use of 25% of the world's pesticides. Some naturally colored cotton which provides promosing research area can also be grown organically [4–6].

2 Naturally Colored Cotton

Cotton (*Gossypium*) pertains to a flowering plant grown from seed. There are 39 different wild species of cotton currently. New world species *G. hirsutum* and *G. barbadense* and old world species *G. arboretum* and *G. herbaceum* are the most popular cultivated types which have commercial importance. All 39 species are unlike in leaf shape, leaf color, flower, seed, lint, lint color and length. The lint color might be white to off-white, different shades of brown (light brown to chocolate and mahogany red) and green (bright or emerald green which quickly fades to a greenish rusty brown). The cotton fiber plant which grows in natural colors along the cultivation is known as naturally colored cotton and their color are obtained without the aid of synthetic or natural dyes. The new awareness on the environment protection has been driving the production of the naturally colored cotton in many different shades such as green, tannin, yellow, brown, pink etc. Many incorrectly used chemicals in dyeing stages may damage human health. Ecologically friendly naturally colored cotton fibers eliminate the dyeing stage in industrial production. Hence the cloth production costs are lowered by leaving out these stages, decreasing water and energy expenditure and the waste quantity to be treated. Furthermore those colorants used for the cloths are very often carcinogenic and harmfull to human health [7, 8].

Naturally colored cotton's (NACOC) origin is South and Central America where it has been cultivated for about 200 years. Native peoples in the former American lands have used these wild cottons for weaving and hand spinning for centuries. It was also reported that Mochica Indians have been growing naturally colored cottons of myriad hues for over the last two millenniums on the northern coast of Peru. There are some records that when the first Spaniards first crossed the Peruvian desert in 1531 they were so surprised to see the extensive fields of colored cotton growing. NACOC fabrics had been collected as the first items of tributes and sold or shipped to Spanish court. Those Indian Textiles were woven on European looms as the most technical sophisticated textile products at the end of 15th century. However, low yields as well as the short fiber length which made them harder to be spun or woven led to some delays for the commercial textile production [9]. In 1982 an entomologist Sally Fox the first breeder in Wickenburg Arizona reintroduced the naturally colored cotton fibers which eliminated the need for dyeing. The colored fibers were developed by hybridization to be long enough for the successful spinning. The achievement in breeding machine spinnable naturally colored cotton led her register a natural colored cotton trademark; Fox Fibre® which included the natural colors of green, coyote, brown, Buffalo brown and Palo Verde green. In 1984, Raymond Bird also started experimental trials for improving the red, green and brown cotton quality. Finally, in 1990, the Bird brothers and C. Harvey Campbell who is a famous agronomist and cotton breeder in California came together and formed BC Cotton Inc. to work with naturally colored cottons. Organic production of naturally colored cotton has also been realized among some of the other countries of the world. There were some ongoing projects related to naturally colored organic cotton in India, Israel and the USA. However, restricted color choices for the consumers with low quality fiber have been the major problems. Although some countries especially Brazil, Peru and China go on producing naturally colored organic cotton, the ratio of organic certificated group is very low [10, 11].

The origin of the brown and green cottons available today on the cotton market is varieties of species *G. hirsutum*. Seeds of the cottons are available in seed banks all over the world [10–13]. Color Colour pigmentation occurs after the opening of the cotton ball. Color of cotton fiber is the results of presence of pigments, which are intermingled with cellulose. This property is a genetically inherited characteristic. Colored cottons may be in different shades of green and brown [14]. After the cotton bolls open, the colors are developed under the sun. Firstly, the cotton fiber is white and after one week, it turns into its original color in compliance with its genotypic constitution. The shades of the naturally colored cotton may differ in accordance with the season of the year and the location because of the climate and soil type [15–20]. It is useful to emphasize that naturally colored cottons have shorter staple length, weaker in staple strength and finer in the diameter when compared with the regular upland cotton. Lower fiber yields have resulted in low usage of color linted cotton variaties in commercial textile applications. However, environmental concerns and desire of the awaked consumers triggered the interest of using naturally colored cotton in textile and apparel sector [21–23]. Naturally colored cotton fiber cultivation and production are also carried out in Turkey with small scale. Moreover, although it is not very common, naturally colored cotton fibers can be used in textile products in Turkey such as blankets etc. Natural colors of such environmental friendly and healthy textile and apparel items exhibit different shades of yellow, brown and green depending on the environmental conditions of crop growing lands, the seed types and therefore more importantly their inherent genes [24, 25]. NACOC has high resistance to insects and diseases which is an important feature that ensures compatibility with organic growth without using pesticide, herbicide etc. It is worth mentioning that naturally colored cotton is not necessarily cultivated in organic methods nevertheless if the ecologically correct methods are conducted, it is a good option as a raw material for ecological fabrics [26].

Today some investigations have been conducted for the improvement of the genetic properties of naturally colored cotton fibers in respect of better yield and better fiber qualities regarding strength, length and micronaire since natural colored cottons are desired to be more competitive against conventional white cottons. NACOC has a cost advantage with the elemination of dyeing process in fabric manufacturing. Additionally, instead of color fading problem which can be encountered in the case of dyed white cotton fibers, the color of the naturally colored cotton fiber becomes stronger after laundering. It has also been declared that clothes made of NACOC have been successful for preventing skin diseases as well as protecting skin from ultraviolet radiation. A significant number of research studies have been carried out on white cottons, while naturally colored cottons were used to be left behind. Nonetheless, when environmental pollution has began to be one of the most urgent and important problem of the world, naturally colored cotton fibers are one of the more preferred options for more sustainable, renewable and ecological textile production [27] (Fig. 1).

2.1 Structure of the Naturally Colored Cotton

Recent investigations of naturally colored cottons have displayed that brown cotton is very similar in morphology to white cotton whereas green cotton is different as it comprises suberin. Suberin, contains mainly biofunctional fatty acids, can normally form a three dimensional network in the existence of glycerol that is found in green cotton fiber nevertheless not in white cotton fiber. Elesini et al. made a study for the confirmation of suberin in green cotton fiber by performing the infrared spectroscopy measurements. According to their results, it was concluded that; the bond $(O_6–H…O)$

Fig. 1 White cotton and naturally colored cotton fiber examples

that is normally found in cellulose I was not observed at the spectrum of green cotton. They also emphasized that the existence of suberin does not affect the structure of the individual crystallites nevertheless obstructs the development of the crystallites in the green cotton fibres [28].

Ahuja et al. investigated the performance of upland colored cotton germplasm. In 1995 and 1996 totally 18 F1 (*Gossypium hirsutum*) colored cotton hybrids were cultivated by crossing six white linted female lines F-505, LRK-516, F-846, HS-6, P-31 and LH-1134 with three colour linted male testers 84 (green linted), 2077 (medium brown linted) and H. Taskant (dark brown linted). The variance analysis results indicated that F1 colored cotton hybrids exhibited higher seed cotton yield, number of bolls/plant, fiber length, fiber strength and heavier bolls when compared with their parents. It was experimentally displayed that the pigments in naturally colored fibers were probably flavonoid compounds, such as flavonone, flavonol and anthocyanidin. It was concluded in this study that, color is controlled by a dominant gene (the genetic factor) and environmental factors which affect mainly the intensity of the color [28–30].

The investigations related to the comparison of the brown and green colored cotton structure with the white and brown cotton structure concluded that green cotton fibers have a different morphology from white and brown. This was explained with the green cotton's secondary cell wall being composed of alternate cellulose-sub ring layers. Green cotton fibers are accepted as different from white fibers in terms of their smoother structure and containing higher organic solvent extractable material. Yatsu et al. carried out a research study where the green fibers (*Gossypium hirsutum* L.) were displayed by electron microscopy (EM) to possess numerous thin concentric rings around the lumen of the cell. EM investigation of both ordinary white and green cotton fibers exhibited that white cotton possesses a very thin (0.02 μm) outer cuticle while the green cotton displayed a series of concentric rings in the cell wall which was observed on the green cotton. Higher magnification of the osmiophilic ring structures of green cotton displayed that each ring consisted of a lamellar pattern

Table 1 Morphological character parameters of natural colored cottons and white cotton [32]

Cotton	Lineer density/dtex	Average length/mm	Curliness
White	1.50	36	17
Brown	1.66	28	22
Bottle green	1.30	25	26
Laurel green	1.39	26	24

characteristic of suberin while white cotton did not reveal any osmiophilic structure in the fiber walls. The cell wall in green cotton usually possesses numerous concentric, osmiophilic rings around the lumen of the cell [31].

When the cell wall of green cotton fiber was magnified, it was observed that the ring structure was made up of at least two parts; the one with a dark, continuous heavy line and the other one with discontinuous lighter series of lines. The extraction of green cotton with $CHCl_3/CH_3OH$ appeared to abolish the dark, heavy lines where the lighter, wispy lines remained. Additionally, the section marked with the letter L revealed the lamellar structure at the higher magnification.

In one of the studies of Zhang et al.'s; The SEM images of naturally colored cotton gave some ideas about the fiber morphology of naturally colored cottons. It was declared that although there are clear cellulose microfibrillars in white cotton fiber surface where those microfibrillars were wider on the surface of brown cottons. The crimpness of the two green cottons was the highest. White cotton was observed as the plumpest with the thickest cell wall. Natural green cotton had the thinnest cell wall compared to others. Table 1 reveals some data indicating the comparison of morphological character parameters of natural colored cottons with the white cotton. Additionally, IR spectra and X-ray diffraction were conducted for the comparison of chemical compositions, crystalline structures and thermal characteristics. The results revealed that crystallinity and crystallite sizes of laurel green cotton had lower values than the bottle green cotton which was attributed to the excess content of suberin in the former one. The crystallinity of brown cotton revealed similar results with the white cotton. Thanks to its higher lignin content, stability of thermal property of bottle green cotton was the best among the others and its decomposition temperature was higher of 30 °C than that of common white cotton [32].

The structure of pigment compositions and radical scavenging activity of the naturally green-colored fiber was investigated in Ma et al.'s study. It was declared in the study that the hue of green cotton was actually yellowish green. Two yellow components in the pigment extact were isolated and named as 22-*O*-caffeoyl-22-hydroxydocosanoic acid glycerol ester and 22-*O*-caffeoyl-22-hydroxydocosanoic acid. The green cotton fiber revealed higher scavenging capacities in comparison with that of common white cotton fiber. Alkaline treatment of green cotton in home laundering led to 80–85% loss of the radical scavenging capacity while only 20–30% loss was obtained by laundering in neutral bath [30].

Ryser et al. investigated the green cotton fibers in the electron microscope. The thin sections of aldehyde osmium fixed fibres revealed concentric, osmiophilic layers

Table 2 General properties of cotton fiber samples [45]

Cotton fiber type	White cotton	American brown cotton	American green cotton
Micronaire index	3.7	3.8	2.6
Maturity ratio	0.86	0.84	0.59
Percentage Maturity	76.8	74.4	50.8
Fineness (dtex)	1.57	1.65	1.38

in the walls which are divided by cellulosic material. The investigators attributed the number of layers to number of days of secondary wall formation suggesting a periodic deposition. They also emphasized about the presence of suberin with the chemical analysis of fibre cell walls and associated the suberin with characterization of the green lint cotton fibers by formation of concentric rings of lamellated lipid material [33].

The researchs related to brown cottons' chemical features and biosynthesis pathway of brown pigments in cotton fiber revealed that the pigments were pertained to proanthocyanidins (PAs). But it should be remembered that PA biosynthesis pathway in brown cotton are still being investigated. PAs known as tannins have different functions as they may be found as the pigments in seed coat or they have responsibility for the protection of the plant against the herbivores or microbes. Additionally, PA's antioxidant and anti-inflammatory properties provide it to become a potential chemopreventive and chemotherapeutic agent for some human diseases, comprising cancers [34–43]. Xiao et al. clarified the details of PA biosynthesis pathway in brown cotton fiber. The gene expression profiles in developing brown and white cotton were compared by using digital gene expression profiling. They concluded that all steps from phenylalanine to PA monomers (flavan-3-ols) were considerably up-regulated in brown fiber when compared to white fiber [44].

Richards et al. investigated some cotton fiber properties of naturally-colored brown and green cottons (Tables 2 and 3). For determining the wax percentage, the cotton samples were extracted with ethanol. The fineness, maturity and tensile properties were evaluated among three cotton types of different colors. The results revealed that green cotton was less mature and as a result of that green cotton was finer than the brown and white cottons. It is stated that although green and brown cottons were weaker and had lower work of rupture (cN/tex) values compared to those of white cotton, the measured properties were declared to be adequate for most textile purposes [45].

As earlier mentioned, the colored cottons have the advantages of being environmental friendly characteristics. These fibers may be used in the textile goods such as shirts, apparels and terry fabrics. It should be stated that there are also complaints about the products being not comfortable owing to their low moisture absorbency. This was attributed to fat, pectin and lignin content that are considered to be hydrophobic in nature. China started a research program and project on colored cotton and they have established four naturally colored cotton research centers

Table 3 Tensile properties of the cotton fiber samples [45]

	White cotton	Brown cotton	Green cotton
Tenacity (cN/tex)	14.33	9.78	9.12
CV (%)	34.8	36.9	41.6
Extension (%)	5.78	4.96	6.18
CV (%)	26.6	23.3	25.3
Specific work of rupture (cN/tex)	0.39	0.20	0.25

Table 4 Compositions of green and white cottons (%) [46]

Substance	Green cotton	White cotton
Fat	4.34	0.6
Lignin	9.34	0.0
Pectin	0.51	1.2

as well as with the growing area of naturally colored cotton exceeding 20,000 ha. Gu conducted a research related to the improvement of the moisture absorbency of naturally self-colored cotton. Warm water and NaOH solution treatment was carried out in this work which aimed to remove the harmfull substances in order to increase the moisture absorbency of the fiber. An improvement for the moisture absorption was obtained after the treatment [46]. The compositions of green and white cotton (in percentage, %) which were utilized in this work were displayed in Table 4.

The researchers emphasized that some hydrophobic substances in the fibre are liable for the water hating behavior of the fiber. They also revealed the microscope image of the naturally green cotton which is a ribbon-like structure twisted at irregular intervals along the fiber length. The cross-sectional view exhibited the lumen which is U-shaped with a central hollow core.

The naturally colored cottons are also promising in terms of their flame resistancy, ultraviolet protection and antibacterial properties. As the cotton is ignited in the presence of oxygen at high temperature necessary for the combustion initiation, the untreated cotton will burn or smolder. Burning is the flaming combustion whereas smoldering is the burning and smoking or wasting away by a slow oxidation without a flame until the carbonaceous materials occurs. Some researchers claimed that colored cottons exhibit better thermal resistance than white cottons. They added that the degradation of naturally colored cottons was observed at about 390 °C where the degradation of white cotton was observed at around 370 °C which was attributed to high amounts of metal [47–50].

Husvedt and Crews exposed the cotton speciments to xenon light and accelerated laundering. When the ultraviolet transmission values were measured, it was observed that naturally-pigmented cottons revealed significantly higher UPF values than conventional cotton (bleached or unbleached). It should be stated that xenon light exposure and laundering caused some fading; however, UPF values of naturally pigmented cotton continued to be sufficiently high for all shades [51].

Li et al. investigated the antibacterial activity of naturally colored cottons. According to their results, brown naturally colored cottons exhibited an outstanding antibacterial activity with a reduction rate of 89.1 and 96.7% when exposed to two species of bacteria named as *Staphylococcus aureus* and *Klebsiella pneumoniae*. However, it should be added that the effect of green NACOC was faint. For investigating the antibacterial mechanism of NACOCs, pigments have been extracted from the naturally colored fibers by using the disk diffusion method. Chemical nature of extracted pigments revealed that the pigment from brown cotton fibers displayed significant inhibition against the two types of bacteria however the resistance capacity of the pigments from green cotton fiber was insignificant. It was also reported that the pigment of brown cotton belonges to tannins whereas the pigment of green cotton was identified as flavonoids. Influence of high temperature treatment on exctracted pigments led to a decrement for the antibacterial activities but it was still at a satisfactory level [52].

In another study related to naturally coloured fibers' structure, fibers and nanofibers were extracted from white and naturally colored cotton fibers (brown, green and rubyy). Studied fibers were investigated with regard to chemical composition (content of cellulose, lignin, hemicellulose and elemental analysis), zeta potential, morphology (SEM for fibers, STEM and AFM for nanofibers), crystallinity (XRD) and thermal stability (TG). Cellulose nanofibers from white and naturally colored cotton fibers were confected by acid hydrolysis. There was not any significant difference in shape and size among the white and naturally colored cottons. STEM and AFM analyses displayed that nanofibers had a length of 85–225 nm and diameter of 18 nm. The nanofibers preserved their original color even after the acid extraction with giving colored suspensions in water. There was not requirement of any pretreatment processes of bleaching etc. The colored nanofiber suspensions were produced to be a possible alternative for the reinforcement materials of polymer. Additionally, researchers concluded that there was a prominent difference between the colored cottons and the white cotton with regard to thermal stability where the colored nanofibers were thermally more stable in isothermal oxidizing conditions at 180 °C than white nanofibers [53].

2.2   Yarn and Fabric Properties Produced from Naturally Colored Cotton Fiber

In some research studies, colored cotton fibers (light and dark cinnamon, champagne and green) were investigated for their features and conversion process from fibers to yarns. Table 5 displays the comparison of selected naturally colored fiber properties with standard white Upland cotton. It was reported that colored cotton comprises higher amounts of hemicellulose and lignin and waxes which provides resistance for natural colored cottons against *Aspergillus niger*. Properties of ring spun yarn and rotor spun yarn produced from natural colored cottons are exhibited in Tables 6 and

Table 5 Comparison of selected fiber properties of naturally colored cotton with standard upland cotton [54]

	Light cinnamon	Dark cinnamon	Champagne	Green	Upland cotton
Strength (g/tex)	26.5	20.6	28.5	23.4	28.5
Elongation (%)	5.2	5.7	4.7	5.5	–
Modulus	7.2	4.8	8.6	5.5	–
Mean length (in)	0.8	0.73	0.92	0.85	0.9
Uniformity (%)	81.0	77.6	82.1	78.5	80.9
Short fiber (%)	10.3	18.6	9.5	12.5	–
Micronaire index	3.9	3.9	4.5	3.5	4.6

7, respectively. Properties of the fabrics made of those blended yarns used as weft in various ratios were investigated in order to evaluate the K/S values after scouring, bleaching and exposure to light. A respectable change in K/S values was obtained especially for the blends with high amount of colored cotton. Light fastness values were lower for the fabrics with high amount of colored cottons but they improved after the treatment with the various chemicals such as tannic acid, aluminum potassium sulfate, copper sulfate, ferrous sulfate etc. [54].

Matusiak et al. investigated the conversion of the naturally colored cotton of Greek origin into yarns as well as into the fabrics composed of these cotton fibers. Ring and rotor spun yarns were selected as the spinning method composed of brown Greek cotton. There was not any major problem encountered with the colored cottons' rotor spinning process. The brown Greek cotton could be processed in rotor mill into Open End yarns of linear densities at the same yarn counts with middle staple white cotton. The plain woven fabrics made of white and brown cotton were produced. Raw and finished fabrics were evaluated in terms of tenacity, elongation, air permeability, thermal resistance, thermal conductivity, thermal absorption and water vapor resistance. It was concluded that the application of the brown cotton in warp preparation and weaving did not affect the industrial processes and the fabric production. It was also declared that the low strength of brown cotton fibers in the warp yarn resulted in a noteworthy decline in the strength of washed fabrics in the warp direction whereas brown cotton did not have any effect on fabric shrinkage. When it comes to the thermal properties of naturally colored cotton fabrics, they exhibited the similar level with those made with fabrics from white cotton [55]. Degirmenci et al. investigated the properties of ring yarns spinned from naturally brown colored cotton as well as the woven and knitted fabrics produced from those yarns. It was concluded from the fastness results that fabrics made from naturally colored cottons exhibited the similar

Table 6 Properties of ring spun yarns produced from the four different colored cottons [54]

		Light cinnamon	Dark cinnamon	Champagne	Green
Skein test	Yarn number (tex)	19.5–37.3	20.7–37.5	19.8–36.8	20–37.4
	CSP (kN m/kg)	2.176–2.6	1.968–2.225	1.968–2.225	2.678–3.046
Single yarn test	Tenacity (kN m/kg)	13.1–15.5	11.3–13.4	13.9–15.7	15.7–17.8
	Elongation (%)	4.9–6.0	5.2–6.5	4.4–5.8	5.7–6.8
	Work to break (kN m/kg)	0.298–0.44	0.292–0.437	0.287–0.422	0.433–0.605
Uster evenness test	Nonuniformity (%CV)	15.5–19.6	16–20.4	16.1–21.1	14.2–18.1
	Thin places/1000 m	23–158	50–318	30–235	0–115
	Thick places/1000 m	290–1.163	315–1.270	343–1.523	143–675
	Neps/1000 m	13–50	3–55	20–215	10–28

Table 7 Properties of rotor spun yarns produced from the four different colored cottons [54]

		Light cinnamon	Dark cinnamon	Champagne	Green
Skein test	Yarn number (tex)	19.5–37	19.7–36.6	20.1–36.6	19.9–36.5
	CSP (kN m/kg)	1.947–2.274	1.689–1.959	2.006–2.266	2.123–2.477
Single yarn test	Tenacity (kN m/kg)	12.2–13.3	10.8–11.8	12.6–13.4	13.4–14.3
	Elongation (%)	4.6–5.0	4.6–5.1	4.6–4.8	5.3–5.8
	Work to break (kN m/kg)	0.276–0.356	0.257–0.336	0.278–0.329	0.364–0.447
Uster evenness test	Non uniformity (%CV)	11.2–12.8	11.5–12.8	11.2–12.8	11.5–12.8
	Thin places/1000 m	0–3	0–10	0–8	0–8
	Thick places/1000 m	3–23	13–15	3–30	3–13
	Neps/1000 m	0–25	5–25	5–20	5–13

fastness properties, apart from light fastness exception. The researchers concluded that not only the spinnability of fiber but also the fabric production from the yarns which are made from naturally colored cotton fibers are successful [22].

Rathinamoorthy and Parthiban carried out a study where they cultivated the naturally colored cotton fibers from the fields of Coimbatore region, Tamil Nadu, India. The total yield of brown color cotton was analyzed for the yarn and fabric development. It was declared that the naturally colored cotton fiber had short upper half mean length, low uniformity index, low fiber strength, good fineness, low elongation (%), good uniformity, and average maturity. After rotor yarns were spun, the yarn characteristics were also analysed. Three different types of fabrics were woven and the produced fabrics were evaluated in terms of tearing strength, crease recovery, stiffness, air permeability, abrasion, and pilling resistance. The fastness properties like color and light as well as duration to washing were also analysed. The researchers concluded that the results were promising and the naturally colored cotton may be developed for the new products such as home textile, casual wears, upholstery fabrics etc. [56]. Sanches et al. carried out a study about the comparison of knitted fabrics produced from naturally colored organic cotton, bamboo, rayon, recycled PET/cotton and recycled polyester fibers. Pilling, rupture pressure, elasticity, elongation, moisture absorption and dimensional tests were applied to fabric samples. The results verified that all raw materials were adequate for clothing manufacturing. Additionally, it was emphasized that the fabrics manufactured with recycled PET/cotton blended yarns displayed greater potential for providing the consumers' needs [57]. Chae et al. conducted a study for investigating the mechanical properties of naturally colored organic cotton in terms of tactile sensory perceptions. Two species, coyote brown and green NACOC fibers, were selected and woven into plain and twill fabrics. The mechanical features were evaluated by Kawabata Handle (KES-FB) system. The researchers concluded that there were significant differences between the shear, surface, compression features and weight of four different naturally organic cotton fabrics [58].

2.3 Influence of Pretreatment and Finishing Processes

It was reported that fabrics made from naturally colored cottons are sensitive to most of the chemicals such as oxidants, reductans, metallic ions, acids and alkali. Additionally, high temperatures in the wet processes and the sunlight radiation in daily use are thought to result in irreversible color change of naturally colored cottons. Scouring is one of the vital processes for NACOC which changes the original color of the fiber. In some studies, it has been declared that scouring treatment changes the original color of NACOC into even a deeper and darker color owing to inner pigments moving outwards [59, 60]. Park et al. investigated the effects of NACOC color, scouring method and age on the visual sensibility of NACOC. Three colors were selected as specimen (ivory, green and coyote brown). The naturally colored cottons (ivory, green and coyote brown) were treated by two chemical scouring tech-

Table 8 Effect of bleaching on naturally colored cottons [20]

Blend percentage of colored cotton	Hunter whiteness index	
	Green cotton	Brown cotton
16.7	84.62	82.57
27.8	83.7	83.21
39.1	83.19	83.86
55.7	74.03	80.46
0[a]	86.11	86.11

[a] 100% white cotton

niques (Na_2CO_3 and NaOH) and two bioscouring techniques (enzyme and boiling water). The samples were evaluated in terms of nine visual sensibility scales (bright-dark; clearmurky; heavy-light; vivid-subdued; warm-cool; freshstale; strong-weak; showy-plain; and luxurious-cheap) among the women raters in two age groups (20s and 30s; 40s and 50s) [61]. Parmar and Sharma investigated the influence of scouring, bleaching and the treatment with various treatments on the weaving fabrics made of yarns with naturally colored fibers blended with the white cotton fiber. The blending process was conducted at the fiber stage in ring spun yarn processing. The researchers revealed that scouring treatment led to lower color depth for the treated fabrics. This was explained with the pigment of leaching out from the colored fiber and distributed between all fibers. The fabric samples were also applied to bleaching (Table 8) and it was observed that both the fabrics with brown cotton fiber type and the green cotton fiber type were very close to the whiteness index compared with the fabrics of 100% white cotton fiber [20].

Williams and Horridge conducted a study related to the evaluation of changes in color of coyote, green and white cottons because of the selected pretreatments used for laundering and dry-cleaning. The researchers concluded that all pre-treatments altered the inherent colors of naturally colored cotton. Green and coyote cotton fabrics became darker with the repeated care cycles. Application of ammonia on a fabric made from naturally colored cotton might be intentionally exposed for developing any desired cotton. But the researchers emphasized not to apply ammonia in any form except developing desired cotton including fabric finishing procedures. Additionally, the usage of chlorine bleach, either concentrated or in solution, was not advised [62]. In another study of Williams, green and brown colored cottons were darker with the repeated washing cycles and dry-cleaning. Color change was obtained with the water temperature and hardness. Compounds in solvents used for removing the stains were found to be effective for the color change [63]. Han et al. made a study which investigated the color changes of Naturally Colored Organic Cotton (NACOC) fibers after scouring, and to assess the human sensory perception for the fibers. Three colors (ivory, coyote-brown, green) of naturally colored cotton fibers were scoured under four different treatments (boiling water, enzyme, sodium carbonate, and sodium hydroxide). The colorimetric properties were measured and

Fig. 2 Naturally brown colored organic cotton fiber (*Gossypium hirsutum* L.), its yarn and its single jersey knitted fabric

color differences (ΔL, Δa, Δb, ΔE) were calculated. Human sensory perception for the NACOC's was evaluated by 27 female participants by using the questionnaire consisting of Human sensory perceptions such as brightness, clearness, lightness and freshness etc. It was observed that values of L* and b* decreased where the value of a* increased after scouring in general and added that the value of ΔE was the highest when treated with alkali solutions among all treatments. Human sensory perception such as brightness, clearness, lightness and freshness mostly diminishes, whereas vividness and strength enhanced [64].

2.4 End Uses and Future Trends

Conventional dyeing process for cotton fibers is water, energy and time consuming process leading to dye effluent which may contribute to environmental problems unless treated in a controlled manner. On the other hand, naturally colored cotton fiber does not need any dyeing or coloration process due to its inherited color characteristics which then resulted in non-usage of synthetic dyes and multiple washing cycles after dyeing process leading to energy, water and time savings with more sustainable future for the world. Producing the textile products by using the naturally colored cotton fibers as the raw materials can be considered as a good way of reducing the environmental pollution. Today there are some efforts in Campina Grande, Paraiba and Brazil where the Coopnatural (Textile Cooperative) purchase the naturally colored cotton produced by small producers in the region. The naturally colored cotton fibers are converted into yarns and fabrics. The families and some associations produce handmade clothes for the "Natural Fashion Brand" by using organic and naturally colored cotton [64]. Although genetically modified seeds affect the cotton production around the world, many groups and indigenous cultures go on fighting for sustainable naturally-colored cottons through seed saving and nurturing their ancient cultivars. Naturally brown colored organic cotton fiber (*Gossypium hirsutum* L.), its yarn and its single jersey knitted fabric are shown in Fig. 2.

3 Conclusion

Increasing consumer demands and environmental concerns resulted with more requirements for sustainable textile materials. The complex nature of the textile products' life cycle has many potential environmental impacts in the context of sustainability approach. Cotton fiber is still the most preferred strategic raw material of textile materials among the world. However it is necessary to lower the chemical residues during the period between cotton cultivation to the cotton end-products in a sustainable production manner. Naturally colored organic cotton fibers have been a promising alternative for their inherent color characteristics which provide more environmentally friendly textile processing with low water, chemical and energy use. Although naturally colored cotton possesses the similar structure with white cotton, it differs from its pigmentation existence at the center of the lumen. Furthermore, the naturally colored cotton fibers are known to be disease and pests resistant which makes them highly tolerant to the organic cultivation methods. The different shades of green and brown are the most common available colors of naturally colored cottons which are believed to be depending on the dominant genetic factors, tannin-suberin and other non-cellulosic materials as well as on the environmental factors. Another amazing feature about the naturally colored cotton is the flame resistance performance with high limited oxygen index owing to the presence of heavy metals in their structure. Additionally, the natural pigments of naturally colored cottons provide good sun protection properties with high UPF values.

As revealed in the early literature studies, the naturally colored organic cotton fibers can easily be utilized for the production of yarns leading to woven and knitted fabrics or also for nonwoven production. However, when compared with the white cotton fiber types, the shorter fiber lengths with low quality lead to restrictions on the naturally colored cottons to be widely used commercially among the world. Cotton breeders should be encouraged for continuing their efforts to expand the number of naturally colored cotton colors and to improve fiber quality and yield by crossing with other genotypes in order to make selections. Naturally colored organic cotton fibers are generally used in handicrafts, t-shirts, blankets, jackets, knitwears, sweaters, socks, towels, shirts, underwear, intimate apparels and other clothing articles, home decorations and furnishings.

In the near future, with the increased awareness and ecological production style, the cultivation, production and consumption of naturally colored organic cotton fibers are expected to increase due to their natural, sustainable, renewable, biodegradable, eco-friendly nature. Therefore, consumers may expect more encounters with the naturally colored organic cotton fibers in their textile products which fulfill and satisfy their lives from many different angles leading to more sustainable world.

References

1. Wakelyn P, Chaudry M (2007) Organic cotton. In: Gordon S, Hsieh YL (ed) Cotton: science and technology. Woodhead Publishing Limited, Cambridge, England, pp 130–174
2. Limitations on Organic Cotton Production (2003) International Cotton Advisory Committee, Washington. Available https://www.icac.org/cotton_info/tis/organic_cotton/documents/2003/e_march.pdf. Accessed 2 Apr 2018
3. Life Cycle Assessment (LCA) of Organic Cotton (2014) Textile exchange. Germany. Available http://farmhub.textileexchange.org/upload/library/Farm%20reports/LCA_of_Organic_Cotton%20Fiber-Full_Report.pdf. Accessed 2 Apr 2018
4. Unsal A Organic Cotton Production and Marketing in Turkey (2013) International cotton advisory committee. USA. Available https://www.icac.org/tis/regional_networks/documents/asian/papers/unsal.pdf. Accessed 2 Apr 2018
5. Devrent N, Palamutçu S (2017) Organic cotton. Paper presented at the international conference on agriculture, Forest Food Sciences and Technologies, Cappadocia/Turkey, 15–17 May 2017
6. Palamutcu S (2017) Sustainable textile technologies. In: Muthu S (eds) Textiles and clothing sustainability. Textile science and clothing technology. Springer, Singapore
7. Horstmann G (1995) Dyeing as a new environmental challenge. J Soc Dyers Col 111:182–184. https://doi.org/10.1111/j.1478-4408.1995.tb01719.x
8. Nascimento ARB, Ramalho FDS et al (2011) Feeding and life history of Alabama argillacea (*Lepidoptera: Noctuidae*) on cotton cultivars producing colored fibers. Ann Entomol Soc Am 104(4):613–619
9. Apodaca JK (1990) Naturally colored cotton: a new niche in the Texas natural fibers market. Working paper series, Bureau of Business Research, paper number 1990-2. Bureau of Business Research, Austin, TX
10. Fox S Naturally coloured cottons. Spinoff, pp 29–31
11. Campbell JR, Harvey C (1994, 1995, 1998) BC cotton, Inc. Personal communication, 1–12–94, 2–16–95 and 6–98
12. Elesini US, Čuden AP, Richards A (2002) Study of the green cotton fibers. Acta Chim Slov 49:815–833
13. Percy R, Kohel R (1999) Qualitative genetics. In: Smith CW, Cothren JT (ed) Cotton: Origin, history, technology, and production. Wiley, New York, pp 319–360
14. Dickerson DK, Lane EF, Rodriguez DF (1999) Naturally colored cotton: resistance to changes in color and durability when refurbished with selected laundry aids. California Agricultural Technology Institute, California State University, Fresno, 1–42
15. Taris Cotton and Oil Seeds Agricultural Sales Cooperatives Union. Available http://www.taris pamuk.com.tr/. Accessed 10 July 2017
16. Organic Cotton Market Report (2014) Organic cotton market report—the textile think tank. Available http://www.thetextilethinktank.org/organic-cotton-report/. Accessed 1 Sep 2017
17. Carvalho L, Farias F, Lima M, Rodrigues J (2014) Inheritance of different fiber colors in cotton (*Gossypium barbadense* L). Crop Breed Appl Biotechnol 14(4):256–260. https://doi.org/10.1590/1984-70332014v14n4n40
18. Chaudry MR (1992) Natural colors of cotton. ICAC Recorder Wash 10(4):3–5 (Technical Information Section
19. Dickerson D, Lane E, Rodrigues D (1999) Naturally coloured cotton: resistance to changes in color and durability when refurbished with selected laundry aid. California State University, Agricultural Technology Institute, Fresno, USA
20. Parmar MS, Sharma S (2002) Development of various colours and shades in naturally coloured cotton fabrics. Indian J Fibre Text Res 27:397–407
21. Rediscovering the value of naturally colored cotton. Available http://www.ideassonline.org/public/pdf/ColoredCotton-ENG.pdf. Accessed 3 Sep 2017
22. Değirmenci Z, Kireççi A, Kaynak H (2010) Investigation of fastness properties of woven and knitted fabrics. Electron J Text Technol 4(2):30–42

23. Cotton Research Institute (2017) Nazilli, Aydın, Turkey
24. Basaravadder AB, Maralappanavar MS (2014) Evaluation of eco-friendly naturally coloured *Gossypium hirsutum* L. cotton genotypes. Int J Plant Sci 9(2):414–419
25. Cotton Association of India (2014) How colourful is the future of naturally coloured cotton. Cotton Statistics and News, Available http://www.cicr.org.in/pdf/Kranthi_art/Coloured_Cotto n_Ap_2014.pdf. Accessed 3 Sep 2017
26. Souza MCM (2000) Produção de algodão orgânico colorido: possibilidades e limitações. Informações Econômicas, Spain 30(6)
27. Republic of Turkey Ministry of Customs and Trade, Directorate General of Cooperatives (2016) Cotton report. Republic of Turkey Ministry of Customs and Trade, Ankara
28. Ahuja SH, Dhayal LS, Monga D (2009) Performance of upland coloured cotton germplasm lines in line x tester crosses. Euphytica 169(3):303–312
29. Xiao Y, Zhang ZS, Yin MH, Luo M, Li XB, Hou L, Pei Y (2007) Cotton flavonoid structural genes related to the pigmentation in brown fibers. Biochem Biophys Res Commun 358:73–78
30. Ma M, Hussain M, Memon H, Zhou W (2016) Structure of Pigment compositions and radical scavenging activity of naturally green-coloured cotton fiber. Cellulose 23(1):955–963
31. Yatsu L, Espelie K, Kollatukudy P (1983) Ultrastructural and chemical evidence that the cell wall of green cotton fiber is suberized. Plant Physiol 73:521–524
32. Zhang L, He J, Wang S (2008) Structure and thermal properties of natural colored cottons and bombax cotton. J Therm Anal Calorim 95(2):653–659
33. Ryser U, Meter H, Holloway P (1983) Identification and localization of suberin in the cell walls of green cotton fibers (*Gossypium hirsutum* L., var. green lint). Protoplasma 117(3):196–205
34. Ioelovich M, Leykin A (2008) Structural investigations of various cotton fibers and cotton celluloses. Bioresour 3(1):170–177
35. Li YJ, Zhang XY, Wang FX, Yang C-L, Liu F et al (2013) A comparative proteomic analysis provides insights into pigment biosynthesis in brown color fiber. J Proteomics 78(14):374–388
36. Feng H, Tian X, Liu Y, Li Y, Zhang X et al (2013) Analysis of flavonoids and the flavonoid structural genes in brown fiber of upland cotton. PLoS One 8(3):e58820
37. Hua S, Yuan S, Shamsi IH, Zhao X, Zhang X et al (2009) A comparison of three isolines of cotton differing in fiber color for yield, quality, and photosynthesis. Crop Sci 49(3):983–989
38. Xiao Y-H, Zhang Z-S, Yin M-H, Luo M, Li X-B et al (2007) Cotton flavonoid structural genes related to the pigmentation in brown fibers. Biochem Biophys Res Commun 358(1):73–78
39. Hua S, Wang X, Yuan S, Shao M, Zhao X et al (2007) Characterization of pigmentation and cellulose synthesis in colored cotton fibers. Crop Sci 47(4):1540–1546
40. He F, Pan Q-H, Shi Y, Duan C-Q (2008) Biosynthesis and genetic regulation of proanthocyanidins in plants. Molecules 13(10):2674–2703
41. Winkel-Shirley B (2001) Flavonoid biosynthesis. A colorful model for genetics, biochemistry, cell biology, and biotechnology. Plant Physiol 126:485–493
42. Santos-Buelga C, Scalbert A (2000) Proanthocyanidins and tannin-like compounds—nature, occurrence, dietary intake and effects on nutrition and health. J Sci Food Agric 80(7):1094–1117
43. Yokozawa T, Cho EJ, Park CH, Kim JH (2012) Protective effect of proanthocyanidin against diabetic oxidative stress. Evid Based Complement Alternat Med 2012. https://doi.org/10.115 5/2012/623879
44. Xiao Y-H, Yan Q, Ding H, Luo M, Hou L, et al (2014) Transcriptome and biochemical analyses revealed a detailed proanthocyanidin biosynthesis pathway in brown cotton fiber. PLOS One 9(1) https://doi.org/10.1371/journal.pone.0086344
45. Richards A, Rowe T, Elesini U (1999) Structure of naturally coloured cottons. J Text Inst 90(4):493–499
46. Gu H (2005) Research on the improvement of the moisture absorbency of naturally self-coloured cotton. J Text Inst 96(4):247–250
47. Lewin M, Sello SB (1975) Technology and test methods of flameproofing of cellulosics. In: Lewin M, Atlas SM, Pearce EM (eds) Flame-retardant polymeric materials. Springer, Boston, MA, pp 19–136

48. Lewin M (1984) Chemical processing of fibers and fabrics. Part B: functional finishes. In: Lewin M, Sello S Handbook of fiber science and technology. Marcel Dekker, New York, pp 1–41
49. Miller B, Martin J, Meiser C (1973) The autoignition of polymers. J Appl Polym Sci 17(2):629–642
50. Parmar MS, Chakraborty M (2001) Thermal and burning behavior of naturally colored cotton. Text Res J 71(12):1099–1102
51. Crews PC, Hustvedt G (2005) The ultraviolet protection factor of naturally-pigmented cotton. J Cotton Sci 9(1):47–55
52. Ma M, Li R, Du Y, Tang Z, Zhou W (2013) Analysis of antibacterial properties of naturally colored cottons. Text Res J 83(5):462–470
53. de Morais TE, Corrêa AC, Manzoli A, de Lima LF, de Oliveira CR, Mattoso LHC (2010) Cellulose nanofibers from white and naturally colored cotton fibers. Cellulose 17(3):595–606
54. Chen C, Cluver K (2010) Biodegradation and mildew resistance of naturally colored cottons. Text Res J 80(20):2188–2194
55. Matusiak M, Kechagia U, Tsaliki E, Frydrych I (2007) Properties of the naturally coloured cotton and its application in the ecological textiles. Fourth World Cotton Research Conference, Lubbock, pp 10–14
56. Rathinamoorthy R, Parthiban M (2017) Colored cotton: Novel eco-friendly textile material for the future. In: Martínez LMT et al (eds) Handbook of ecomaterials. pp 1–21
57. Sanches RA, Takamune K, Guimarães B, Alonso R et al (2014) Wearbility Analysis of knited fabrics produced with colored organic cotton. Bamboo rayon, corn, recycled pet/cotton and recycled pet/polyester. Am Int J Contemp Res 4(4):28–37
58. Chae Y, Lee M, Cho G (2011) Mechanical properties and tactile sensation of naturally colored organic cotton fabrics. Fibers Polym 12(8):1042–1047
59. Church J, Cho G (2007) Determining the psychoacoustic parameters that affect subjective sensation of fabric sounds at given sound pressures. Text Res J 77(1):29–37
60. Kang S, Epps H (2008) Effect of scouring on the colour of naturally-coloured cotton and mechanism of colour change. AATCC Rev 8(7):38–43
61. Park J, Chang Y, Hong W, Lee M, Chae A, Cho G, You H (2012) Effect of colour, scouring method and age on the visual sensibility of naturally colored organic cotton (NACOC). Hum Factors and Ergon Manuf Servi Ind 24(3):1–10
62. Williams BL, Horridge P (1996) Effects of selected laundering and drycleaning pretreatments on the colors of naturally colored cotton. Fam Consum Sci Res J 25(2):137–158
63. Williams BLM (1994) Fox fibre naturally colored cotton, green and brown (coyote): resistance to changes in color when exposed to selected stains and fabric care chemicals. Ph.D. thesis. Available https://ttu-ir.tdl.org/ttu-ir/bitstream/handle/2346/61180/31295008692369.pdf?sequence=1. Accessed 3 Sep 2017
64. Han A, Chae Y, Lee M, Cho G (2011) Effect of color changes of naturally colored organic cotton fibers on human sensory perception. Fibers Polym 12(7):939

Organic Cotton and Cotton Fiber Production in Turkey, Recent Developments

Gizem Karakan Günaydin, Arzu Yavas, Ozan Avinc, Ali Serkan Soydan,
Sema Palamutcu, M. Koray Şimşek, Halil Dündar, Mehmet Demirtaş,
Nazife Özkan and M. Niyazi Kıvılcım

Abstract Cotton has been used for many years in many different regions of the
world. It is a strategic fiber owing to its wide usage leading to high employment
opportunities in textile sector. Turkey has a considerable contribution on the world
textile and apparel industry with its cotton fiber growing capacity and textile and
clothing manufacturing capacity. Cotton cultivation areas in the world are located
between the north parallels of 32–36 which reveal the warm climate features. Turkey
locates in the north board of world cotton cultivation area. In Turkey, cotton farming
is carried out mainly in four regions: Aegean, Çukurova, Southeastern Anatolia and
Antalya where climate and grown cotton properties of each region differs. "Why
organic cotton in Turkey" is an important question which has many reasonable and
satisfying answers. Turkey has well established organic cotton regions and cotton
yields are considerably high in the country. In this chapter, organic and conventionally
grown cotton fibers are handled with a broad perspective in terms of cotton fiber
cultivation and recent development about these fiber types in Turkey. Firstly, organic
cotton and organic cotton fiber cultivation in Turkey, organic cotton growing regions
in Turkey, limitations for the organic cotton markets, lack of Information on cost
of production, marketing and future trends will be reviewed and discussed in detail.
Moreover, general cultivation in lands and cotton fiber yield in Turkey are given in
detail and information about the encountered diseases and pests during the cotton
fiber cultivation are explained.

Keywords Organic cotton · Turkish cotton · Organic agriculture · Cultivation
Conventional cotton fiber · Disease · Fiber yield · Pest

G. K. Günaydin
Buldan Vocational School, Pamukkale University, Buldan, Denizli, Turkey

A. Yavas · O. Avinc · A. S. Soydan · S. Palamutcu (✉)
Textiles Engineering Department, Pamukkale University, Denizli 20016, Turkey
e-mail: spalamut@pau.edu.tr

M. Koray Şimşek · H. Dündar · M. Demirtaş · N. Özkan · M. Niyazi Kıvılcım
Cotton Research Institute, Nazilli, Aydın, Turkey

© Springer Nature Singapore Pte Ltd. 2019
M. A. Gardetti and S. S. Muthu (eds.), *Organic Cotton*, Textile Science
and Clothing Technology, https://doi.org/10.1007/978-981-10-8782-0_5

1 Introduction

Cotton fiber utilization dates back to ancient times. The cotton textile pieces belong to 2700 BC found in the excavations of Mohenjo Daro in Indus river valley, within the boundaries of Pakistan today and fabric samples belong to 2500 and 1700 BC, found during the excavations of 'Ancon Chillon' ruins, located in the mid coast region of Peru, may be declared as the first evidence that the cotton has been used in the World since over 5 thousand years [1, 2].

Cotton is an important fiber since it has a wide usage and provides high possibilities for the employment. Cotton industry has a vital economic importance in terms of the growing countries. Cotton fiber is used in textile sector whereas its seed is used for oil-feed industry and its linter is used for the regenerated cellulosic fiber industry and paper industry. In the recent years, there has been an increasing trend in cotton oil usage since it is a good alternative for petroleum. Cotton fiber is mainly cultivated to meet the textile sector's requirements.

90% of cotton agriculture is conducted in the in northern hemisphere and the land ratio is about the 2.5% of agricultural land of the world. Furthermore, cotton farming provides employment in many sectors such as ginning, storing, utilization of farming equipment, etc. Increase in population and the rising living standards lead the higher demand for natural cotton biofiber each year [3].

2 World Cotton Production and Developments

Cotton has a strategic importance as being the important raw material of textile industry. In order to produce a quality product, using a uniform raw material has a vital importance. However, this is not easy as it seems for the natural fiber such as cotton which is difficult to be cultivated with the same quality and minimum variation. Cotton as the primary consumable natural fiber type has been cultivated in different areas of the world. Main cotton fiber growing countries in the world are listed in Table 1 where Turkey has the 8th place after India, China, USA, Pakistan, Brazil, Australia and Uzbekistan [4]. Table 2 reveals cotton yields of countries by years. According the data of International Cotton Advisory Committee, for the year 2015–2016; the world average has cotton yield of 69,595 (kg/ha) whereas Turkey has cotton yield of 1475 (kg/ha) leading to 3rd place in the rankings (Table 2).

India, China, USA, Pakistan, Uzbekistan, Brazil, Turkmenistan, Turkey and Australia are the main countries, which constitutes 80% of the cotton cultivation region of the world [3–5].

Main cotton exporting countries are ordered as USA, India, Brazil, Australia, Uzbekistan and Burkina Faso, whereas main importing countries are Bangladesh, China, Vietnam, Turkey, Pakistan, and India for the year of 2015/16 [3, 4, 6] (Tables 3 and 4). According to the data of ICAC; 69–75% of the imported cotton and 81–82% of the exported cotton is supported by the 10 countries mentioned in Table 3 and 4,

Table 1 Major cotton fiber growing countries in the world (1000 mt) [6]

	Country	2012–2013	2013–2014	2014–2015	2015–2016	2016–2017[a]
1	India	6290	6766	6562	5746	5775
2	China	7300	6950	6550	4988	4871
3	United States	3770	2811	3553	2806	3738
4	Pakistan	2002	2076	2305	1537	1663
5	Brazil	1310	1734	1563	1289	1485
6	Uzbekistan	1000	910	885	832	789
7	Australia	1018	885	528	629	960
8	Turkey	745	688	720	640	703
9	Turkmenistan	370	337	333	315	292
10	Burkina Faso	264	270	298	244	285
Total of ten countries		24,069	23,427	23,297	19,026	20,561	21,994
World total		26,783	26,173	26,222	21,328	22,985	24,604
Rate of ten countries (%)		89.87	89.51	88.85	89.21	89.45	89.39

[a]Estimated

Table 2 Cotton yields of countries by years (kg/ha) [6]

Rank	Country	2012–2013	2013–2014	2014–2015	2015–2016	2016–2017[a]	2017–2018[b]
1	Australia	2303	2258	2779	2330	1722	1737
2	Israel	1786	1897	2020	1891	1761	1892
3	Turkey	1527	1546	1565	1475	1674	1685
4	Mexico	1511	1625	1577	1523	1590	1559
5	Brazil	1465	1546	1601	1350	1580	1561
6	China	1390	1448	1520	1484	1558	1558
7	South Africa	777	1172	1205	1208	839	1106
8	Greece	887	1190	1007	908	1009	1028
9	Syrian Arab Republic	1100	976	976	879	983	954
10	United States	999	921	939	859	972	908
World average yield		784.30	799.66	773.46	694.95	776.97	774.25

[a]Estimated
[b]Projection

respectively. In the cotton planting season of 2015–2016, the highest cotton import was provided from China as 959,000 tons (Table 3) whereas the highest cotton export was supplied by USA as 1,993,000 tonnes (Table 4).

3 Cotton Cultivation in Turkey

Turkey has a considerable contribution on world textile and apparel industry since the country is one of the leading cotton fiber grower in the world. Turkey has about 2.8% share in the world's cotton fiber production according to the yield of 2015/16 (Table 1). Since 1990s, rising of grown cotton production rate (%) was observed in Turkey [7]. Cotton fiber has a significant role in the national economy in Turkey. Not only its fiber is used in textile sector but also cotton seed serves as a raw material in oil industry, and its cake may be used as a feed by stock breeders. Cotton farming and cotton fiber processing also provides a large channel for employment [7]. So it is useful to analyze the properties and the variations of cotton in Turkey over the years for determining the Turkish cotton standards.

Turkey comes as the 3rd country in terms of cotton fiber yield after Australia and Israel in the year of 2015/16 (Table 2). Although the cotton fiber yield is very high in Turkey, there is an increment in cotton import of Turkey from 381,000 tons to 755,000 tons between the years of 2000 and 2015. This result was attributed to the insufficient domestic production of cotton fiber and to the trend of new textile investments in Turkey. USA, Brazil and Greece provide 80% of cotton lint import of Turkey and Turkmenistan, Uzbekistan, Argentina, India, Tajikistan and Egypt supply the rest of it. Table 5 reveals the harvested area (ha), production (tonnes) and yield (kg/ha) of Turkey by the years.

Figure 1 displays the cotton production (raw-lint) between the years of 1991–2016 in Turkey and Fig. 2 displays the Turkey cotton harvesting area (in decares) and cotton production (in tonnes) [9]. It can be stated that there are some fluctuations on the total cotton fiber yield and cultivation area among Turkish cotton grower society between years of 1991–2015. The highest cotton harvesting area is recorded at 1995 as 756,694 hectare and then total cotton harvesting area is decreased constantly and according to the records of 2016, only about 400,000 hectare of cotton cultivating area were utilized. It is also revealed in the Fig. 2 that annual cotton production does not show such high fluctuations which is as result of increasing amount of cotton yield among Turkish cotton growers.

Turkey's cotton export and import amounts between the years of 2000–2016 are revealed in Fig. 3. According to the data obtained from 2017 data Turkish Statistical Institute; Turkey cotton import was 566,784 (tonnes) in year 2000 and it reached its maximum value in the year 2007 as 946,213. Turkey imports an average of 800,000–900,000 tons of cotton per year. In 2016, Turkey imported 821,217 tons of cotton fiber for Turkey textile production. Although cotton export value was reported as 27,515 (tonnes) in year 2000, this value has increased every year and reached to 76,131 (tonnes) in 2016.

Table 3 World cotton import ($\times 1000$ tonnes) [6]

Rank	Country	2012–2013	2013–2014	2014–2015	2015–2016	2016–2017[a]	2017–2018[b]	2012–2016
1	China	4426	3075	1804	959	1057	1071	12.392
2	Bangladesh	1055	1112	1183	1378	1412	1545	7.685
3	Viet Nam	517	687	934	1001	1239	1301	5.679
4	Turkey	803	924	800	918	704	714	4.863
5	Pakistan	411	247	166	585	581	374	2.364
6	India	258	147	267	234	475	329	1.710
7	Mexico	215	223	211	222	230	236	1.337
8	Brazil	14	32	5	20	50	22	143
9	South Africa	17	19	19	16	17	10	98
10	Greece	4	3	4	6	5	5	27
Total of ten countries		7.720	6469	5393	5339	5770	5607	
World total		10.214	8860	7787	7569	7954	7870	
Rate of ten countries		75.58	73.01	69.26	70.54	72.54	71.25	

[a]Estimated
[b]Projection

Table 4 World cotton export (×1000 tonnes) [6]

Rank	Country	2012–2013	2013–2014	2014–2015	2015–2016	2016–2017[a]	2017–2018[b]	2012–2016
1	United States	2836	2293	2449	1993	3157	2873	12.765
2	India	1690	2015	914	1258	909	927	6.023
3	Australia	1343	1057	520	616	704	759	3.656
4	Brazil	938	485	851	939	610	650	3.535
5	Uzbekistan	690	615	550	500	337	370	2.372
6	Burkina Faso	247	270	252	272	247	261	1.302
7	Turkmenistan	175	354	325	273	178	161	1.291
8	Greece	238	280	254	209	209	210	1.162
9	Pakistan	97	111	96	50	24	26	307
10	Argentina	55	43	84	48	60	47	282
Total of ten countries		8309	7523	6295	6158	6435	6284	
World total		10,050	9034	7743	7555	7875	7720	
Rate of ten countries		82.68	83.27	81.30	81.51	81.71	81.40	

[a]Estimated
[b]Projection

Table 5 Cotton production of Turkey for the last 12 seasons

Cotton production			
Crop year	Harvested area (ha)	Production (tonnes)	Yield (kg/ha)
2005–2006	546,880	863,700	1580
2006–2007	590,700	976,540	1650
2007–2008	530,253	867,716	1640
2008–2009	495,000	673,400	1360
2009–2010	420,000	638,250	1520
2010–2011	480.650	816,705	1700
2011–2012	542,000	954,600	1760
2012–2013	488,496	858,400	1760
2013–2014	450,890	877,500	1950
2014–2015	468,143	846,000	1810
2015–2016	434,013	738,000	1700
2016–2017[a]	415,000	756,000	1800

[a]Estimated [8]

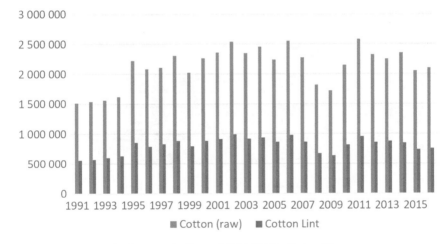

Fig. 1 Cotton production between the Years of 1991–2016 in Turkey [9]

It is observed that the best cotton cultivating areas in the world are coordinated between the north parallels of 32–36 which reveal the warm climate features. Turkey locates in the north board of world cotton cultivation area. As in the other branches of agriculture, the main objective in cotton farming is to get more and better quality products from the unit area. Genetic potential, environmental conditions and breeding techniques are major parameters that influence the amount and quality of the cotton crop. Climate and soil are the most important factors that limit the cotton cultivation

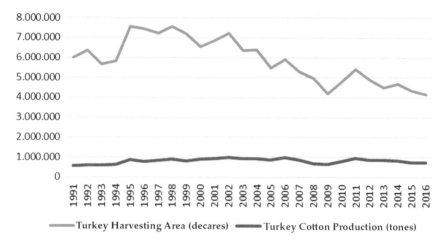

Fig. 2 Turkey harvesting area (decares) and Turkey cotton production (tonnes) [9]

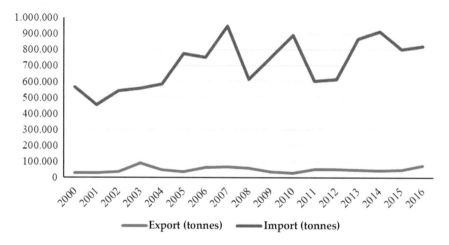

Fig. 3 Turkey cotton import and export (tonnes) [9]

areas. Turkey provides conventional and organic cotton types. When the transgenic cotton is concerned, transgenic cotton cultivation is not allowed in Turkey.

Various diseases, insects and pests can damage cotton fiber production in Turkey. Some of the encountered diseases and the causes are listed as follows [10]:

- **Cotton seedling disease** caused by *Pythium* spp., *Rhizoctonia solani*, *Fusarium* spp., *Thielaviopsis basicola*, *Verticillium* spp., *Alternaria* spp., *Macrophomina* spp.
- **Verticillium wilt on cotton** *caused by Verticillium dahliae Kleb*
- **Cornered leaf spot disease** caused by *Xanthomonas axonopodis* pv. *Malvacearum* (Smith) Vauterin.

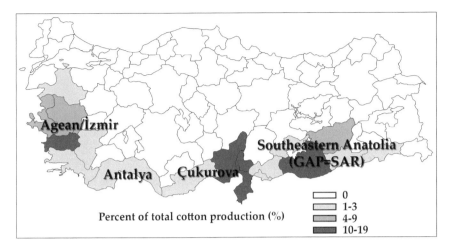

Fig. 4 Cotton production regions in Turkey [11]

Important encountered insects and pests in Turkey may be listed as follows;

- Aphid Species problem caused by *Aphis gossypii Glov*.
- Trips tabaci problem caused by *Thrips tabaci Lind. (Thys., Thripidae)*
- Cotton leaf fleas problem caused by *Empoasca decipiens (Paoli) Asymmetrasca decedens (Paoli) (Hom., Cicadellidae)*
- Carmine Spider Mite problem caused by *Tetranychus urticae Koch. (Acarina., Tetranychidae)*
- Green mite problem Caused by *Heliothis armigera HBN*
- Pinky mite problem caused by *Pectinophora gossypiella Saund. (Lep., Gelechi-idae)*
- White fly problem caused by *Bemicia tabaci Genn. (Hom., Aleyrodidae)*.

In Turkey, cotton cultivation is carried out primary in four regions: The Aegean region, Çukurova region, Southeastern Anatolia region and Antalya region (Fig. 4). For the present, the Aegean region has the greatest share in cotton output of Turkey. Quality of the cotton fiber of this region is also the highest comparing to the other regions outputs. In Çukurova area, output is subject to significant fluctuations and in general, it exhibits a declining trend. Çukurova Region was an important area in 1980s in terms of Turkey's cotton production for both cultivating areas and total output. After the 1990s, the production in the region was in decrement trend because of both the cost increases owing to the raid against increasing number of vermins and disease and also the reduction of planting areas because of the pricing policy in practice. It could also be said that the loss of farm workers and forcing the Southeastern Anatolia Project (SAP) are the other factors which resulted in reduction in production [4].

Cotton production by Regions (Table 6) reveals that fiber production increased until year of 2011/2012 then decreased in the subsequent years. The total value was 737,000 tonnes at the end of 2015/2016. Another remarkable result about Table 6 is increased cotton fiber production in Aegean and Mediterranean regions until 2011/2012 and decreased in the subsequent years. But this was not valid for the Southeastern Anatolia Region (SAR) where the cotton cultivation area increased regarding to the increase of irrigated fields based on the national Giant Anatolian Project GAP project. "GAP" (SAR: Southeastern Anatolia Project) is a big project which dates back to 1970s. The water and land resources were developed with the hydroelectric power plants on the Euphrates and Tigris Rivers. "GAP" project includes 22 dams, 19 hydroelectric power plants and 1.8 million hectares of irrigation area. The new investments were expedited with the irrigation of Harran Plain in 1995. The ratio of cotton cultivation areas increased in SAR Region and in 2011 this region had the 61% of the total cultivation area of Turkey. In 2015 marketing year, SAR Region has grown 58% of the total cotton production in Turkey [12] (Table 7).

The cotton fiber yield according to regions is displayed in Table 8. Seed cotton yield was observed as higher in Mediterranean and Aegean regions than SAR Regions. This result was attributed to better climate conditions in these regions and the sufficient knowledge of the cultivators. In 80 years, seed cotton yield increased up to six times. Seed cotton yield is a specific characteristic which depends on environmental conditions as well as the crop management activities and also genetics of the seed. Liu et al. declared that the yield depends on 48% genetics, 28% crop management and 24% interaction of crop management and genetics [13]. Krieg also concluded that 70% of cotton-yield variation influences from the environmental conditions every year and 30% of cotton-yield variation influences from crop management system [14].

There are some studies related to variations in Turkish cotton properties depending on the cultivating area as well as the conditions such as climate, ground etc. These studies emphasize that cotton fiber cultivated in different locations of Turkey reveal different fiber characteristics in terms of fiber lengths, micronaire, strength, neps and the Spinning Consistency Index (SCI). Some of the researchers also analyzed the influence of SAR Project on the cotton quality cultivated in South East Area of Turkey and concluded that some cotton groups from this location revealed more superior fiber characteristics compared to other regions [15–17]. The two groups of cotton one of which was cultivated in Aegean Region of Turkey and the other in Urfa (South-East) Region of Turkey was measured with the HVI (High Volume Instrument) for this study to reveal the cotton fiber's characteristic property varying according to the regions (Table 9). HVI and AFIS (Advanced Fibre Investigation System) instruments are commonly used for evaluating the quality of cotton fiber in the bale. HVI was designed to measure the fiber properties from a bundle of fiber while AFIS instrument was designed to measure the single fiber. Typical HVI measurements include fiber length, length uniformity, bundle tenacity, elongation, micronaire, color, and trash content while AFIS instrument measures length, fineness maturity, circularity, short fiber content, immature fiber content, neps/g and percentage of dust and trash [18].

Table 6 Cotton production by regions in Turkey (1000 mt) [8]

Cotton growing regions	1934–1938	1974–1976	1993–1994	2001–2002	2011–2012	2012–2013	2013–2014	2014–2015	2015–2016
Mediterranean	35	248	164	220	250	193	188	165	145
Aegean	18	217	272	291	167	151	176	184	165
SAR[a]		53	144	411	536	513	512	495	427
Means of Turkey	53	518	580	922	953	857	876	844	737

[a] SAR Southeastern Anatolia project

Table 7 Cotton acreage area by regions in Turkey (1000 ha) [8]

Cotton growing regions	1934–1938	1974–1976	1993–1994	2001–2002	2011–2012	2012–2013	2013–2014	2014–2015	2015–2016
Mediterranean	195	333	154	152	130	102	89	85	78
Aegean	40	236	237	236	97	82	83	94	92
SAR[a]		85	148	298	314	302	279	289	265
Total	237	696	559	697	541	486	450	468	435

[a] SAR Southeastern Anatolia project

Table 8 Cotton fiber yield by regions in Turkey (kg/ha) [8]

Cotton growing regions	1934–1938	1974–1976	1993–1994	2001–2002	2011–2012	2012–2013	2013–2014	2014–2015	2015–2016
Mediterranean	200	820	1060	1350	1930	1880	2120	1960	1870
Aegean	430	920	1150	1240	1720	1830	2140	1970	1800
SAR[a]		630	970	1390	1710	1700	1840	1720	1620
Total	300	790	1060	1330	1790	1800	2030	1880	1760

[a] SAR Southeastern Anatolia project

Table 9 HVI results of Aegean and Urfa cotton

Cotton type	SCI	Mic.	UHML	SFI	Strength (gr/tex)	Elg.	Neps (gr)	Rd	(+b)	%Rh
Aegean Cotton	149	4.93	30.22	6.45	34.17	7.28	118	71.66	8.46	7.42
Urfa cotton	143	5.05	29.49	6.9	33.5	6.4	96	72.0	8.7	6.3

SCI Spinning consistency index, *Mic.* Micronaire, *SFI* Short fiber index, *Str* Strength *Elg.* Elongation, *UHML* Upper half mean length in inches, *UI* Uniformity index *Rd* Reflectance degree, *(+b)* Yellowness of cotton fibre, *%Rh* Relative humidity

3.1 Better Cotton Grown in Turkey

"The Better Cotton Initiative (BCI) is the largest cotton sustainability schedule in the world". Better cotton is a way of strategy which brings together farmers, ginners, traders, spinners, mills, cut & sew, manufacturers, retailers, brands and grassroots organizations in a unique global community which aims to develop "Better Cotton" as a sustainable mainstream material. This program includes supporting the farmers by procuring training and learning opportunities for them to adopt more environmentally, socially and economically sustainable manufacturing practices. BCI licensed farmers produce cotton in an environmental awareness with the efforts of minimizing the effects of fertilizers, pesticides and caring for waters, soil health and natural habitats [19]. Estimations for BCI cotton display that 30% of the world's total cotton manufacture will be BCI cotton in 2020 [20]. In Turkey "BCI's IP IPUD (Good Cotton Practices Association)" was founded in 2013 for the application of Better Grown Cotton strategy on Turkish cotton agriculture. IPUD's mission is defined as "to improve cotton production in Turkey for the benefit of cotton producers and the regions where cotton is grown and for the future of the sector."

IPUD (Good Cotton Practices Association) conducts field visits and holds training events to increase awareness of BCI Farmers about topical issues. In 2016, it was built on those efforts by developing a comprehensive decent work training program, in partnership with the Fair Labor Association (FLA), comprising a broad range of decent work topics. According to Better Cotton initiative 2015/2016 Turkey Harvest report [20]; 441 farmers in Turkey earned a Better Cotton license in Turkey. The report also emphasized that many BCI farmers who had planted at the optimal time were less affected from the spring, autumn rains as well as from the high temperatures which lead to increasing the average yield of BCI farmers in comparison to other farmers. It was declared in the same report that BCI Farmers had applied, 12%, on average, less pesticide ingredient than other farmers (Fig. 5) [21].

3.2 Research Institutes/Organizations Involved in Cotton Research in Turkey

The Ministry of Food, Agriculture and Livestock-(MFAL) is the principal state organization involved in agriculture and rural development in Turkey. The Ministry coordinates and implements the agricultural R&D activities through the General Directorate of Agricultural Research and Policy (GDARP). The General Directorate of Agricultural Research and Policy is the headquarter of the national agricultural research system and it is liable for determining national research strategy, constituting research priorities and assigning available financial resources to the programs and assisting the government in developing and improving agricultural policy. Other major contributors in cotton research are universities, the Scientific and Technological Research Council-TUBITAK, Scientific Research Project Coordina-

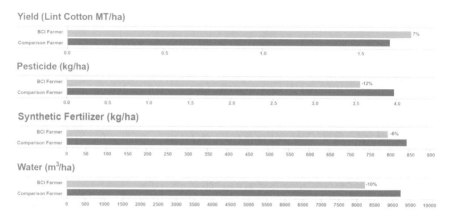

Fig. 5 Comparison of BCI Turkish farmer and conventional Turkish farmer in terms of yield, usage of pesticide, synthetic fertilizer, and water [21]

tion Department of Universities and the private sector companies. The Ministry has many research institutes, but the Cotton Research Station in Nazilli-Aydın-Turkey is the primary mono crop multidisciplinary research institute [8].

Variety Approval and Seed Supply Turkey has been a member of the International Union for the Protection of New Varieties of Plants (UPOV), and variety development and adoption follows the UPOV guidelines. Certificated delinted planting cotton fiber seed is utilized in Turkey. The breeder seed trials are done at least two locations in every season, or one location in two subsequent seasons to complete trials. The best breeder lines are then sent along with the detailed report to the Seed Certification and Approving Center, Ankara for approval. The committee in the Approval Center decides according to the results of two years (at least four locations) cotton fiber varieties.

4   Organic Cotton

Organic agriculture is a production management system which enhances biodiversity as well as soil biological activity. This production is based on the applications of maintaining and enhancing the ecological harmony [21]. "Organic Cotton" fiber can be defined as "more sustainable" than the conventional cotton" fiber which is an environmentally preferable product. The organic cotton proponents support the idea of "conventional cotton is not an environmentally responsibly produced crop". Since the conventional cotton production has a disadvantage of overuse or misuse of pesticide/crop protection products, it has adverse effects on the environment. Additionally conventionally grown cotton fiber/fabrics/apparel has chemical residues on the cotton which may cause cancer and some other health related troubles.

Table 10 Organic cotton growing rank in the world yield [23]

	2007–2008	2008–2009	2009–2010	2010–2011	2011–2012	2012–2013
1	India	India	India	India	India	India
2	Syria	Turkey	Syria	Syria	Turkey	China
3	Turkey	Syria	Turkey	China	China	Turkey
4	China	Tanzania	China	Turkey	Tanzania	Tanzania
5	Tanzania	China	USA	USA	USA	USA
6	USA	USA	Tanzania	Tanzania	Mali	Burkina Faso
7	Uganda	Uganda	Uganda	Egypt	Peru	Egypt
8	Peru	Peru	Peru	Mali	Uganda	Mali
9	Egypt	Egypt	Egypt	Kyrgyzstan	Egypt	Uganda
10	Burkina Faso	Burkina Faso	Mali	Peru	Burkina Faso	Peru

The certifying companies are well-known among the all parties who have released their own organic cotton production standards. Although the Technical Information Section of the International Cotton Advisory Board gives information related to Organic Cotton by publishing many articles related to subject, data from some countries cannot be reached & updated [22, 23]. It is thought that production has not increased too much except India, Turkey and USA. Table 10 lists the organic cotton growing countries according to the total organic cotton outputs for the seasons from 2007–2008 to 2012–2013.

5 Organic Cotton Production

Current production practices in conventional and organic cotton are similar in some aspects and not in other aspects regarding the operations. In the condition that all conventional methods are followed, it is ineligible to certificate the crop as organic cotton. It is certain that organic cotton production does not consume most synthetically compounded chemicals (fertilizers, insecticides, herbicides, growth regulators and defoliants) which are suggested for only conventional cotton production. Both organic and conventional cotton productions are followed with crop rotation. The crop rotation reduces the weed problem which may be caused from organic production conditions. Synthetically compounded Conventional cotton production chemicals (fertilizers, insecticides, herbicides, growth regulators and defoliants etc.) should not be used in the field for at least three years before organic cotton production can begin. Hand picking is the most used method among the world which does not differ between organic and conventional cotton harvesting (Fig. 6).

If it is succeeded to remove the fiber from the cotton boll instead of the whole boll and other contaminants from the workers and the field, hand picking may result

Fig. 6 The organic cotton fiber and its hand picking from the cotton field in Turkey

with the cleaner cotton. Green leaves should be removed before harvesting since low leaf drop slows the harvest and also leads to an increment in ginning costs. Hence, it is recommended not to wait and harvest as soon as possible after the bolls have opened.

5.1 Some Limitations for the Organic Cotton Markets

Organic cotton cultivation has not highly extended to the most of the countries as it was expected. There are some reasons for this situation. Since 1990, nineteen countries began to produce organic cotton however the most of them had to stop organic cotton production due to economical limitations. Organic Trade Association took the responsibility for determining the main problems about the organic cotton production. However, the results were not satisfying because of the growers being un-reluctant to share the information about the organic cotton. ICAC's survey emphasized main problems including the cost of production of organic cotton versus that of conventional cotton and the price premium on organic cotton. The fiber council declared that weed management without using herbicide, defoliation and insect control are the main problems for organic control producers. Defoliation problems were frequently encountered with as a serious problem when the organic cotton was picked by the machines. Some varieties may not perform well and reach the expected yield level without synthetic fertilizers and insecticides even under optimum conditions. The other limitations for organic cotton cultivation may be ordered as follows [21, 22].

Fertilizers: Nutrient requires alteration from minimum to maximum for N, P and K during the course of cotton development. Synthetic fertilizers are used in order to meet nutrient requirements of cotton plant. Instead of using synthetic fertilizers, organic fertilization and green manuring may be preferred to provide the required nutrient supply. P and K can stay in the soil and they can be sufficiently achievable whereas the availability of N to the level of inorganic fertilization may not be achieved. Some new alternatives for synthetic fertilizers which do not lead insufficient nutrient supply should be investigated.

Pest Control: Cotton may be attacked by a wide variety of insects including cotton bollworm, pink bollworm, tobacco bollworm, spotted bollworm, etc. Synthetic chemical insecticides are prohibited in organic cotton production which is one of the biggest differences in organic farming. However, "natural" chemicals like sulphur dust may be applied. It should not be forgotten that the first two years of cultivation may be difficult for the transition from synthetic insecticides. The third year is more convenient for the natural balances.

Cotton plant which is vulnerable to insects may be exposed to insect attacks during organic growing conditions. Biological controls may prevent the insect pressure without using any insecticides. Cotton plant may cope with the early losses however after a certain time the plant cannot make up with a loss. Since cotton growing has a definite time for cutting-out period, all precautions to prevent the plant from insects should be taken to save the maximum number of buds, flowers and bolls at the beginning.

Production Control: In the conventional cotton growing, technology package which involves the best usage of inputs and production practices are free and delivered directly by the extension service to farmers' doorsteps. The main advices or technology package for the producer includes the ways to obtain maximum yield, guidance for variety selection, planting time, soil preparation, elimination of weeds, irrigation, insect control, all the way to picking and storage of seed cotton until it is sold. Organic cotton farmers require advice, which is essential to guide them for such a risking investment. It is wrong to assume that organic cotton is more basic than conventional cotton cultivation just because of the elimination of fertilizers, pesticides and other agrochemicals. In fact, it is more challenging to cultivate cotton without agrochemicals [24].

Lack of Information on Cost of Organic Cotton Production

It is a very hesitating decision for the producers whether to grow organic cotton. When the fertilizers and insecticides were introduced in the world countries, there was in an increment in cotton yields. But the current trends display that average yields are not ascending and that the cost of conventional cotton production seems inflated which affect the economics of producing cotton. In that case higher costs without yield increase force farmers to give up the cotton production or continue cultivating it by subsidize production of governments under increased costs. Generally it is believed that organic production has lower costs of production. However, reducing expenditures for planting area may be determined by the effect on yield. Lower total costs/ha should not be accepted as indicator for the low cost of lint. The elimination of prohibited synthetic agrochemicals is generally believed to change the average yield however there is not a comparison data of organic cotton versus that of conventional

Table 11 Opportunities in organic cotton growing business [25]

1	Organic weed controls
2	Cotton seed varieties better suited to growing conditions
3	Marketing efforts to increase consumer demand for U.S.-grown organic cotton
4	Market development to encourage better gate pricing
5	Improved awareness of the GOTS label within the U.S. market
6	Continued improvements to crop insurance
7	Streamlined administrative process for the organic grower
8	Tax credits, certification cost share and other financial incentives to encourage organic production

cotton production in terms of production cost. This result lead the farmers become reluctant to adopt organic cotton production [22].

Price Premium

Generally organic cotton farmers had been expecting to be awarded with the premium price however it was observed that organic producers have even been penalized for producing lower grade cotton because of the boll-worm damage. The collected data over the ten years revealed that there was not any sign related to average premium or discount on organic cotton versus conventional cotton. It is thought that without price premiums, it is impossible to make profit for the producers because of the discernments in the yield [22, 25].

Need for Alternate Inputs

The elimination of synthetic fertilizers and pesticides abridge the plant being protected against nutrient starvation and protection against insect and pests, unless alternative systems having the same quick and effective properties are developed. Manual and mechanical weed control systems exist but in the large fields they are not effective. Alternatives to insecticides and fertilizers are not fast enough. There is not a definite message that whether organic cotton production gives lower or higher yields compared to conventional practices which does not encourage the new producers to adopt organic production.

Marketing

Marketing which is the most sensitive aspect of organic cotton production provides the expansion of market linkages between the cotton farmers and the international buyers. Organic cotton has certainly some advantages as well as some limitations. The opportunities according to organic cotton growers are revealed in Table 11 [25]

Certification

Certification is a way of approval which a company, process or product is in accordance with the requirements of a particular standard. Certification provides a kind of legal protection for purchase assurance with ensuring that environmental or social improvements. Each buyer is advised to check their suppliers for Scope Certificate and Transaction Certificates to help ensure the benefits to the farmer and the environment [26].

5.2 Organic Cotton in Turkey

"Why organic cotton in Turkey" is an important question which has many reasonable and satisfying answers. Turkey has well established organic cotton regions and cotton yields are considerably high in the country. It is possible to cultivate long staple organic textile for fine textile production as well as shorter staple options. Turkey is grower and also the manufacturer of organic cotton who converts the raw material to textile products with the advantage of proximity to European customers who are the major consumers. Additionally, it cannot be ignored that Turkish manufacturers have long time experience in textile manufacturing and specialized in material innovation, design and fashion as well as efficient production and manufacturing. Turkish government has a confirming strategy for the organic cotton for the progression of Corporate Social Responsibility and developed environmental and social management in textile production, in compliance with the Sustainable Development Goals (SDG) [26, 27].

Organic cotton production began in Kahramanmaraş (in the Eastern Mediterranean region, Turkey in 1989–1990. The project was named as Good Food Foundation and was followed by a second multinational project started in Salihli (Manisa, Turkey) in the Aegean region by a German based company. Turkey significantly increased its organic cotton production during 1999–2000 and 2000–2001 [22]. Turkey was considered as the largest producer between the seasons of 2000 to 2007. Although, since 2007, organic production in India increased very much, Turkey remains established as one of the three organic cotton producers among the world.

South Eastern Anatolia Project (GAP) which is the largest production project took place in Turkey, lead an increment in the organic cotton production. According to report of Turkey Organic Cotton Sourcing Guide (2013), the production fell over the past years but begins to rise again after 2013. Now Turkey ranks second after India, producing approximately 16,000 MT (approximately 12%) of the world's organic cotton production amount.

5.3 Organic Cotton Growing Regions in Turkey

Turkish organic cotton is mainly cultivated in the Aegean in the west of Turkey and in the SAR (Southeastern Anatolia Project) region. Aegean part provides the 20–30%

Fig. 7 Organic cotton growing regions in Turkey [11]

of Turkish organic cotton and the rest is provided from SAR region. Both locations have the advantage of being home to biggest organic textile manufacturers.

Since 2000, Cotton growing (including organic) became more popular in SAR region which is mainly due to lower production and labor cost with the effect of GAP project (Turkish version of SAR: Southeastern Anatolia Project) by the Southeastern Anatolia Project Regional Development Administration. GAP project created employment in the region hence number of seasonal workers leaving the area from east to west reduced [28] (Fig. 7). As it is seen in the Fig. 7 and Table 10 the cotton growing area in the south east of Turkey has flourished and widened during the last two decades.

There are some big cooperatives in Turkey which have enough experience for producing crops under the organic conditions (TARIS, Nazilli Cotton Research Institute Republic of Turkey Ministry of Food Agriculture and Livestock). Nazilli Cotton Research Institute is a big foundation under the Republic of Turkey Ministry of Food Agriculture and Livestock which has been investigating on conserving of genetic stock, developing source material, breeding, collection and evaluation of data on cotton in National Scale since 1934 [29]. TARIS (Cotton and Oil Seeds Agricultural Sales Cooperatives Union)'s history extends to the years of 1910s. The foundation is listed in "the biggest 500 organizations of Turkey" which contributes to the Turkish economy with the gin industry through cotton ginneries with a capacity of 656,000 tons/year [30]. Both of the foundations have research centers for the better cotton improvements and new solutions that have low impact on the environment.

5.4   Future Trends for Turkey

Today organic cotton is preferred in more than 20 countries with more than 2000 brands. As the demand for the consumers raised in developed countries, worldwide known brands have been consuming and selling more organic cotton products day by day. Turkey has the advantage of full chain of organic processes from raw material to the finish product [25, 30]. This makes it a promising country which can create its own brand strategy by using its high quality organic cotton converted into full chain of organic cotton products such as t-shirts, baby wear, towels and home textiles.

It should be reminded that brand with the organic cotton certificate require special attention during ginning, baling and also in the packing process. Packing material must be also selected among organic cotton based textile fabrics. Spinning process needs the consideration of elimination of any previous fiber contamination risk to the organic cotton blend.

Special care is required in the sizing process where only appropriate sizing agents needs to be utilized. Appropriate sizing agents should be applied with a special care in weaving preparation. Weaving and knitting stages are regular as it is in conventional cotton manufacturing steps. In the case of wet processing such as the dyeing and finishing processes, only allowed chemicals should be utilized.

Organic textile manufacture steps are also controlled and certified from the raw cotton stage all the way through spinning, knitting/weaving, wet processing, and cut/sew stages as it is in the organic cotton fiber cultivation. The most frequently implemented certification program in organic textiles are GOTS (Global Organic Textile Standards), OEKOTEX 100 [13], and Nordic Swan Ecolabel [14].

6   Conclusion

Cotton fiber cultivation and its processing stages to manufacture any kind of textile product is an established and well-known manufacturing sector in Turkey. Historically the lands of modern Turkish Republic had been known as Asia Minor and the land has very rich ancient civilization collection which were involved with cotton fiber cultivation, hand craft manufacturing of cotton fiber based textile and ancient trade roads assigned back to the 3000 B.C.

Currently farmers in Turkey has been cultivating cotton fiber in their own lands that is estimated about 500,000 hectares for the crop year of 2017. Total cotton output and total land share used for cotton plantation has been increasing in Turkey, where it was about 9% share of the total agricultural land in 2015, 15% share in 2016 and 20% share in 2017. According to the annual report of the ICAC for 2015–2016 season; Turkey is at the 9th place in the world cotton planted area list, at the 2nd place in the world highest cotton yield list, at the 7th place in the world cotton fiber output list, at the 4th place in the cotton consumer country list, and at the 5th place in the cotton exporting country list.

Turkish cotton farmers and agricultural research institutes have been focused on the cotton fiber back after about 15 years. Textile and apparel manufacturing industries are consumed almost all of the cotton fiber output of the Turkey and also requires high amount of cotton fiber import from other cotton growing countries. Cotton fiber consumption in the world textile markets are always priority over the enlarging synthetic fiber market. Besides the modern cotton farming methods, organic cotton farming and naturally colored cotton has also rising interest in the world niche markets. Turkey has also become one of the main organic cotton fiber and naturally colored cotton fiber growing country with willingly farmers and research institutes. Turkish government has a confirming strategy for the organic cotton for the progression of Corporate Social Responsibility and improved environmental and social management in textile production, in compliance with the Sustainable Development Goals. As a result of this, Turkey ranks second after India, producing approximately 16,000 MT (approximately 12%) of the world's organic cotton production amount.

Nazilli Cotton Research Institute (Aydın, Turkey) is a big foundation under the Republic of Turkey Ministry of Food Agriculture and Livestock which has been investigating on conserving of genetic stock, developing source material, breeding, collection and evaluation of data on cotton in National Scale since 1934. Although there are some small amount of scientific studies in the universities regarding the transgenic cotton in Turkey, transgenic cotton fiber production, planting and cultivation are not carried out in Turkey. Since transgenic cotton cultivation is not allowed in Turkey. The quality of Turkish cotton fibers is being tried to be improved with the past, current and upcoming researches and efforts which have been undertaken by the Nazilli Cotton Research Institute, the Turkish universities, private sector companies and with their bilateral cooperation and collaborations. In Turkey, these mentioned institutions and companies are working to improve the properties and quality of different types of cotton fiber genotypes (white, off-white and naturally colored) by using the conventional plant breeding techniques such as a selection breeding and crossing leading to the production of better and more productive superior varieties of cotton fibers. All these efforts about cotton fibers are so important and will add up to the more sustainable World.

References

1. Gulati A, Turner A (1928) A note on the early history of cotton. Indian Central Cotton Committee Technical Laboratory
2. Lee A (2015) Cotton as world crop. Cotton agronomy monograph. American Society of Agronomy, Wisconsin, pp 6–24
3. Çopur O (2017) Cotton production in Turkey. In: 52. HRVATSKI I 12. MEĐUNARODNI SIMPOZIJ AGRONOMA, Dubrovnic/Croatia
4. Republic of Turkey Ministry of Customs and Trade, Directorate General of Cooperatives (2016) Cotton report, Ankara
5. International Cotton Advisory Committee (ICAC) (2016) February monthly report, Washington, DC, USA
6. ICAC (2017) https://icac.gen10.net/statistics/index [Online]. Date of access: 08 July 2017

7. Gül M, Koç B, Dağıstan E, Akpınar M, Parlakay O (2009) Determination of technical efficiency in cotton growing farms in Turkey: a case study of Cukurova region. Afr J Agric Res 4(10):944–949
8. Turkish Statistical Institute (2017) http://www.turkstat.gov.tr/PreTabloArama.do?metod=search&araType=vt [Online]. Date of access: 03 Sept 2017
9. Tuik (Turkish Statistical Institute) (2017) http://www.tuik.gov.tr/PreTablo.do?alt_id=1001 [Online]. Date of access: 04 Aug 2017
10. Kirkpatrick T, Rockroth C (2001) Compendium of cotton diseases, 2nd edn. American Phytopathological Society (APS Press). ISBN: 0890542791
11. Republic of Turkey, Ministry of Economy (2016) 75th Plenary Meeting of the ICAC, Islamabad/Pakistan
12. Southeastern Anatolia Project Regional Development Administration (2017) http://www.gap.gov.tr/istatistiki-veriler-sayfa-63.html [Online]. Date of access: 29 Aug 2017
13. Liu S, Constable G, Reid P, Stiller W, Cullis B (2013) The interaction between breeding and crop management in improved cotton yield. Field Crops Res 148:49–60
14. Krieg D (1997) Genetic and environmental factors affecting productivity of cotton. In: Proceedings of the Beltwide cotton conference, New Orleans
15. Cengiz F, Göktepe F (2006) An investigation of the variation in Turkish cotton fibre properties in years 2002 and 2003. J Text Appar 16:271–275
16. Göktepe F, Göktepe Ö, Şahin B (2002) The impact of giant anatolian project (GAP) of Turkey on cotton production pattern and fibre properties. Fibres Text East Europe 10(2):12–16
17. Gülyasar L, Göktepe F (2000) Spinnability of Turkish cottons with an emphasis on the spinning consistency index (SCI) and count strength product (CSP) parameters. In: The inter-regional cooperative research network on cotton, a joint workshop and meeting of the all working groups, Adana
18. Uster Technologies (2011) https://www.uster.com/en/instruments/fiber-testing/uster-hvi/ [Online]. Date of access: 05 Sept 2017
19. http://bettercotton.org/about-better-cotton/ [Online]. Date of access: 28 Aug 2017
20. Better Cotton Initiative (2016) Annual report, Genève-Switzerland
21. Better Cotton Inititative (2015) Better cotton initiative 2015/2016 Turkey harvesting report
22. Wakenly P, Chudry M (2007) Organic cotton. In: Gordon S, Hsieh Y (eds), Cotton: science and technology. Woodhead Publishing Limited, Cambridge, pp 130–174
23. International Cotton Advisory Committee (2003) Limitations on organic cotton production, Washington
24. http://farmhub.textileexchange.org/upload/library/Farm%20reports/LCA_of_Organic_Cotton%20Fiber-Full_Report.pdf [Online]. Date of access: 6 Sept 2017
25. Unsal A https://www.icac.org/tis/regional_networks/documents/asian/papers/unsal.pdf [Online]. Date of access: 30 Aug 2017
26. Devrent N, Palamutçu S (2017) Organic cotton. In: International conference on agriculture, forest food sciences and technologies, Cappadocia/Turkey
27. Organic Cotton Sourcing Guide—Turkey (2013) http://farmhub.textileexchange.org/farm-library/special-reports
28. TUGEM (2017) http://www.tarim.gov.tr/Konular/Bitkisel-Uretim/Organik-Tarim [Online]. Date of access: 14 July 2017
29. Southeastern Anatolia Project Regional Development Administration (2017) http://www.gap.gov.tr/ [Online]. Date of access: 01 Sept 2017
30. Republic of Turkey Ministry of Food, Agriculture And Livestock (2017). Available: http://arastirma.tarim.gov.tr/pamuk/Menus/45/Our-Tasks [Online]. Date of access: 10 July 2017

Organic Cotton and Its Environmental Impacts

P. Senthil Kumar and P. R. Yaashikaa

Abstract Organic cotton is delivered utilizing an extensive variety of cultivating practices and cotton generation is exceedingly specialized and extremely troublesome. All techniques for creating cotton have impacts that are not really earth well disposed but rather are important to deliver the product. Organic cotton creation has a progression of social and financial dangers, particularly for little ranchers in creating nations. Therapeutic expenses and a powerlessness to work are an extreme monetary weight on influenced families. The over the top utilization of synthetic manures and pesticides in monoculture causes soil debasement, lessening its supplement and water maintenance limit. As a result, agriculturists confront declining yields and need to build generation inputs. Danger of cotton preparing: Most of these chemicals, for example, substantial metals, formaldehyde, azo colours, benzidine or chlorine fade, cause natural contamination by the factories' waste water and numerous can be found as build-ups in the completed item. Notwithstanding, developing cotton utilizing organic generation practices is not appropriate for all nations and all agriculturists. Organic production is not really any more or any less ecologically well disposed than current ordinary cotton generation. For the textile procurer, there is no contrast between routinely developed cotton and organically developed cotton as to pesticide build-ups. Developing natural cotton is more requesting and more costly than developing cotton routinely. Organic generation can be a challenge if bug weights are high however, with duty and experience, it could give value premiums to cultivators willing to address the difficulties.

Keywords Organic cotton · Environmental impacts · Textile · Certification Lifecycle assessment · Toxic chemicals

P. Senthil Kumar (✉) · P. R. Yaashikaa
Department of Chemical Engineering, SSN College of Engineering,
Chennai 603110, India
e-mail: senthilkumarp@ssn.edu.in

© Springer Nature Singapore Pte Ltd. 2019
M. A. Gardetti and S. S. Muthu (eds.), *Organic Cotton*, Textile Science
and Clothing Technology, https://doi.org/10.1007/978-981-10-8782-0_6

1 Introduction

Cotton is a standout amongst the most widely recognized fibres in the apparel sector and most important wellspring of pay for the greater part of smallholder agriculturists. To check negative effects of cotton development in India, for example, natural debasement and money related reliance because of high information costs, organic cotton development is being advanced by non-legislative associations in the nation. Natural organic agribusiness empowers smallholder farmers in a nation to enhance their occupation by giving access to prepare and arrange in gatherings. Green Revolution, given by the presentation of chemically synthesised manures and pesticides, high-yielding assortments, agriculture apparatus and water system innovation, effectively added to the modernization of agriculture and to expanding profitability since the late 1960s [1].

Cotton developed without the utilization of any artificial compound chemicals such as pesticides, plant development controllers, defoliants and manures is viewed as 'organic' cotton. Though, chemicals considered natural can be utilized as a part of the manufacturing of organic cotton and natural composts. *Bacillus thuringiensis* (Bt), a naturally found soil bacterium, can be utilized as a characteristic bug spray in organic farming. Bt is the bacterium, that creates the insect poisons that researchers use to produce genes for bug safe biotech cottons. Biotech cottons, containing Bt qualities, are not permitted to be utilized for the generation of organic cotton—the general reason being that the procedure is not natural [2, 3]. The production of cotton utilizing organic cultivating strategies looks to keep up soil quality and to utilize materials and practices that upgrade the environmental balance of normal frameworks and coordinate the parts of the cultivating framework into a biological entirety. There are a few rules that describe ensured organic cultivating: biodiversity, reconciliation, maintainability, regular plant nourishment, and managing natural pest [4].

As a matter of first importance, ordinary cotton generation has brought about an assortment of serious natural emergencies, for example, decreased soil quality, loss of biodiversity, and hazardous medical issues to the individuals who have been exhibited continuously to the dangerous chemicals utilized as a part of pesticides. Furthermore, pesticides are a major piece of the regular cotton generation. In addition, cotton crops, traditional and also natural, require huge measures of water. Three-year transitional periods from ordinary to organic cotton production i.e. end of engineered chemicals and composts are required for certification [5]. Certified organic cotton is created as per specific nation level or universal organic farming standards, coordinating environmental procedures, keeping up nearby biodiversity, and staying away from the utilization of dangerous and relentless manufactured pesticides and composts and additionally hereditarily modified seeds. Free certification offices licensed by associations that keep up strict controls over certifiers are in charge of ensuring organic cotton as indicated by nation specific norms. To keep up certification through getting ready for the final product, the natural organic cotton must be kept separate from noncertified cotton and be traceable from the land to the finished product [6].

2 Why Natural Cotton?

Supporters of organic cotton and the individuals who showcase organic cotton items advance the recognition that traditional cotton isn't an environment dependably delivered harvest. Traditional cotton creation significantly abuses pesticides/crop protection items that adversely affect the earth and horticultural labourers and ordinarily developed cotton fiber/textures/clothing has synthetic deposits on the cotton that can cause disease, skin problems, and other wellbeing related issues to buyers. Defenders additionally show that organic cotton is a more economical approach. Natural cotton generation isn't identical to sustainable—either natural or conventional cotton manufacturing practices might be supportable [7]. Supporters of organic cotton additionally feel that natural cotton is a more supportable way to deal with cotton generation and propose that organic production is comparable to feasible. Natural cotton generation isn't equal to economical—either natural or regular cotton creation practices might be reasonable or unsustainable. Natural cotton starts from natural farming and is developed without the utilization of any engineered rural chemicals, for example, manures or pesticides. Its creation likewise advances and improves biodiversity and organic cycles. Natural cotton is right now being developed effectively in numerous nations; the biggest makers are the United States, Turkey, India and China [8].

Natural cotton creation does not utilize artificially intensified chemicals but rather can utilize 'common' chemicals like Bt splashes and other organic control specialists in pest administration, and can utilize natural corrosive based foliar showers and nitrogen and zinc sulfate in collect planning. These normal chemicals utilized as a part of gather arrangement are not as powerful for leaf drop, which can prompt slower collecting, lessened review of the cotton and expanded cost of ginning, when cotton is machine picked. Hand picking does not require defoliation whether it is natural or traditional creation however there can at present be green leaf waste close by picked cotton that can influence review, if the plant isn't defoliated before gathering.

Organic cotton is presently developed in excess of 12 nations yet speak to just a small amount of the aggregate cotton creation worldwide. Cotton which is delivered by cultivating framework under Good Agriculture Practices as indicated by set up models without the utilization of GMO seed, engineered chemical manures, pesticides, development controller and defoliant. Organic cultivating assembles assorted farming framework, renews and keeps up soil fruitfulness, adjust in normal natural life form and advance a solid situation [9].

3 Production of Organic Cotton

Organic cotton production does not just mean supplanting manufactured manures and pesticides with natural ones. Natural development techniques are constructing more in light of learning of agronomic procedures than input-based traditional generation is. The foundational approach expects to build up a differing and adjusted cultivating

Fig. 1 Organic cotton. *Source* https://mygreenmattress.files.wordpress.com/2015/03/cotton-field.
jpg

biological community which in a perfect world incorporates a wide range of harvests and farm exercises. Farmers need to finish a two-year transformation period to change their generation framework from traditional to natural [10]. A fundamental component of natural creation is the cautious choice of assortments adjusted to neighbourhood conditions as far as atmosphere, soil and to bugs and diseases. Soil quality administration and product nourishment depend on crop broadening and natural sources of input, for example, fertilizer and composts. Pest administration measures centre basically around bug prevention and the incitement of an adjusted agro ecosystem through yield revolution, blended development, trap crops, and the utilization of characteristic pesticides when bug invasion transcends the monetary limit [11].

At each assembling stage, organic apparel producers don't include oil scours, formaldehyde, antiwrinkling operators, chlorine dyes, or other unauthentic materials. Natural replacement, for example, normal turning oils that biodegrade effectively are utilized to encourage turning; potato starch is utilized for measuring; hydrogen peroxide is utilized for dying; and characteristic vegetable and mineral inks and fasteners are utilized for imprinting on natural cotton texture. These common options are utilized to lessen and dispense with the poisonous outcomes found in customary cotton texture fabricating [12] (Fig. 1).

4 Certification for Organic Cotton

Certification is an essential for an item to be sold as 'organic' cotton. Affirmation gives an assurance that a particular arrangement of gauges has been followed in the production of the organic cotton. Cotton was first affirmed as natural in Turkey. Presently there are many private natural principles worldwide for every single organic item. European Union (EU) directions, International Federation of Organic Agriculture Movements (IFOAM) benchmarks, and the US National Organic Standards (NOP) have defined natural cultivating enactment and guidelines all through the world. Guaranteeing organizations build up their own models yet all are basically equivalent. Organic cotton makers need to resolve to take after the guidelines set by the guaranteeing associations/organizations, which incorporates confirmation through field visits by autonomous outsiders. The confirming office must be certifying, perceived by purchasers, and the framework must be free and straightforward. The charge for certification must be sufficiently low so it doesn't add essentially to the cost of generation.

Mark claims, for example, 'green', 'ecologically well disposed', 'clean' or 'characteristic' are utilized by a few producers however are not endorsed for use as claims by the natural labelling necessities in the International Organization for Standardization (ISO) norms ISO 14020 (ISO, 2000) and ISO 14021 (ISO, 1999), the US Federal Trade Commission 'Green Guides' (2008a) and the UK Department for Environment, Food and Rural Affairs Green Claims Code (2009). The claims are general cases with no administration or official definition. At the point when such claims are utilized to offer items they are viewed excessively ambiguous as significant, making it impossible to the shopper and purchaser recognition and substantiation issues may emerge. Conversely, the agricultural product 'organic cotton' has a specific definition, i.e. cotton created without the utilization of engineered chemicals and following the natural generation anticipate current natural cultivating strategies, that is secured by obligatory government controls [13].

5 Certifiers for Organic Cotton Goods

The following are few certifiers that may be found on certified organic cotton goods:

- Indian National Programme for Organic Production (NPOP)
- Global Organic Textile Standards (GOTS)
- European Organic Regulations (EU 2092/91)
- USDA National Organic Program (NOP)
- Export Certificates for Japan (JAS Equivalent)
- Quebec Organic Reference Standard (CAAQ)
- IOFAM Basic Standards.

6   Economic, Social and Environmental Impacts of Organic Cotton

- **Environmental Impacts**

The principle environmental effects were the change of soil conditions, diversification of culturing pattern, decreased arrival of dangerous agrochemicals to the earth, and conservation of biodiversity. Land utilized for cotton generation had likewise expanded, in spite of the fact that it is confusing whether this has had positive or negative effects. Enhanced soil conditions and diversified culturing pattern were critical changes revealed by agriculturists. As indicated by agriculturists' perceptions, soil conditions on fields enhanced through organic development. Some portrayed a difference in their soil structure in the field. As indicated by agriculturists, these impacts were because of changes in their cultivating rehearses, for example, the utilization of fertilizer or fluid compost, organic fertilizers, and crop rotation. Diminished arrival of dangerous agrochemicals and safeguarding of biodiversity are readditional impacts revealed by farmers [14]. The vast majority of the talk with farmers had already utilized high amounts of synthetic pesticides. The arrival of dangerous agrochemicals to nature is kept away from, which affects the earth, on beneficial bugs and different organisms. Numerous farmers likewise portrayed a lower pest population in their fields. An expansion of land utilized for cotton generation through the presentation of organic cultivating. The greater part of the farmers revealed they now utilize more land to develop cotton, with most detailing a 10% expansion in the land region under cotton development. The expansion in land for cotton triggers a land-utilize change. Land that was regularly utilized for touching, in upland fields and remote tough regions, is currently developed with cotton. This expands the opposition for land and assets as for food production. Upgraded development of cotton can in this way move asset utilize far from sustenance yield to cash crop development [15]. Water is an important factor while considering the economic impacts of cotton. Cotton does not require more water for cultivation usually. In India, cotton is mostly cultivated in drier areas. Average water required for cotton cultivation is 14,000 l/kg. Depending on climate and growth period 700–1300 mm of water is needed for cotton production. Organic water consumes about 182 l/kg of total irrigated water while conventional cotton production requires 2210 l/kg of water which is more than that of organic cotton production.

- **Economic Impacts**

The primary economic effects stated and depicted because of transformation to natural organic production were reduction in input costs, less reliance on cash specialists, less demanding access to seeds and higher market control. The less demanding access to seeds and higher market control are not immediate consequences of a transformation to natural cultivating. Diminished input costs and less reliance on cash banks were critical economic effects because of the transformation to organic. Natural organic cultivating diminished the generation costs for the agriculturists, often significantly, as substance inputs were supplanted without anyone else's input made natural manures and bio-pesticides. The diminishment of generation costs was likewise seen in past examinations on the presentation of natural cultivating rehearses in cotton creation. Less demanding access to seeds and higher market control are not immediate outcomes of transformation to natural cultivation. Through seed banks and cooperatives, farmers have better access to nourishment yield and cotton seeds. This guarantees seed sway for agriculturists and a lower reliance on business sectors and costly hybrid seeds. Expanded wage and more differing pay sources are extra economic effects. This brought about an expansion in the income. Numerous portrayed a higher profit edge and a superior financial circumstance through natural organic cultivating. This was because of various reasons, for example, reduction in input costs, higher costs because of the organic cost premium, more regions under cotton development, extra incomes from diversified food crop generation, and expanded or balanced out cotton creation. One vital influencing factor as for expanded income is yield improvement.

- **Social Impacts**

The social effects because of organic development incorporate better living conditions, wellbeing enhancements, expanded sustenance security, interests in youngsters' instruction and enhanced way of life. Different effects incorporate strengthening and limit working through preparing and organization building, incorporating an expansion in agriculturists' information and fortified groups. As one negative social effect, an increased in work load was noticed. Better living conditions came about because of the presentation of natural cotton cultivating. Wellbeing upgrades were one of the significant changes of transformation to natural cotton generation. Through natural generation, an enhanced nourishment security circumstance happened in the family units of individuals. Accessibility of food expanded as organic cultivating pattern supports food development and higher sustenance crop yields because of better cultivating strategies. Access to nourishment enhanced as the higher wage because of natural cultivating. Strengthening and limit building were other social effects coming about because of the presentation of natural cultivating. As organic farming is somewhat knowledge in-depth, an abnormal state of data is required. During cultivation of organic cotton, as a negative social impact, workload was found to be high [16].

7   Lifecycle Assessment of Organic Cotton

Life Cycle Assessment (LCA) is a perceived tool to gauge and evaluate the natural weights of creation frameworks or items and to find change possibilities. The technique permits objective and logical assessment of the asset necessities of an item and its potential effect on nature during each period of its generation, utilize, and transfer. The LCA approach was connected in a substantial scale consider embraced by the cotton business to assess the ecological effect of ordinary cotton cultivating practices and material creation frameworks [17]. The named contemplate has furnished a strong standard with state-of-the-art Life Cycle Inventory (LCI) information for assessing ordinary cotton items and has started enthusiasm among partners along the whole material production network in examination of ecological execution of their supply chains. To extend and widen the comprehension of natural effects of natural cotton fiber creation, Textile Exchange dispatched PE INTERNATIONAL to play out a Life Cycle Assessment. As the clothing business has turned out to be increasingly dynamic in supportability activities along their supply chains, cotton—one of the essential crude materials—has picked up a great deal of consideration. The objective of organic cultivation is to build up an economical administration framework for agriculture that regards nature's frameworks and cycles, adds to biodiversity and guarantees dependable utilization of vitality and regular assets. Besides, it means to deliver a scope of sustenance and rural items while not spoiling the earth, people', plants' and creatures' wellbeing and welfare. So as to satisfy these goals, organic cultivating and preparing must act as per the measures of natural creation characterized by a few national and worldwide experts.

Cotton development incorporates four primary undertakings: field planning, planting, field activities, and collecting. Under the aggregate term field activities, water system, weed and pest control, and treatment are incorporated. These activities expend vitality (power and fuel), require inputs (seeds, manures, and water) and create squanders and outflows—all of which frame some portion of the present framework. Inside the extent of natural agribusiness pest and weed control are generally preventive as opposed to coordinate. The Life Cycle Impact Assessment comes about are hard to explain as independent markers. Just on the off chance that they are put into point of view to information and logical writing, conclusions on the ecological execution of natural cotton can be drawn.

The product display empowers the figuring of different natural effect types. The effect classes portray potential impacts of the generation procedure on the earth. Environmental change is picked as effect class as environmental change is regarded to be a standout amongst the most squeezing natural issues of our circumstances and there is a huge open and institutional enthusiasm for the concern. Acidification causing acid rain and eutrophication, otherwise called over fertilization, are picked on the grounds that they are firmly associated with air, soil, and water quality and are important ecological effects. The significance of water use in farming frameworks is obvious. This is the reason an ecological evaluation of water utilize is particularly vital in evaluation of agricultural items.

8 Result of LCA

The ozone harming substances for the generation of organic cotton accounts up to CO_2 counterparts. Field outflows allude to gases radiated from soils because of rural action. Basically, these emanations get from microbial supplement change forms in the soil. Because of such change forms, a small amount of the accessible aggregate nitrogen ends up inorganic nitrous oxide, otherwise called laughing gas, with a dangerous atmospheric warming potential very nearly 300 times higher than carbon dioxide. As cotton is a brief purchaser product, this carbon dioxide is discharged later toward the end of the product, with the goal that it is just incidentally put away. This is the reason the uptake of carbon isn't considered in the effect evaluation. Accepting a lifetime of the fiber of 10 years, 10% of the carbon deposited in the item could be credited as a decrease in an Earth-wide warming potential.

At a first look, the commitment investigation draws a comparative picture to that of Global Warming: Field discharges contribute the most took after by ginning and apparatus. While CO_2 emanations add to Global Warming Potential, the parallel arrivals of SO_2 and nitrogen oxides increment Acidification Potential. The effect of field emanations is commanded by alkali while nitrogen oxides and sulfur dioxide discharges impact different procedures inside the generation chain of organic cotton fiber. Sulfur dioxide generation is reliant on the sort of petroleum product utilized and nitrogen oxides rely upon states of the ignition procedure.

Eutrophication in agriculture can be essentially affected by soil disintegration. Through soil disintegration, supplements are expelled from the developed framework by means of water and soil and prompt the treatment of neighbouring water bodies and soil frameworks. Eutrophication Potential is estimated in phosphate-reciprocals and is impacted fundamentally by P-and N containing mixes. Low soil disintegration rates could be accepted resulting in generally low ecological potential [18].

9 Difference Between Organic Cotton Production and Conventional Cotton Production

Organic agricultural practices advances a more secure and more advantageous condition by lessening the introduction of harmful chemicals utilized as a part of traditional cotton production. Natural cultivating techniques enhance soil quality, safeguard the condition and ensure air, water and sustenance supply from destructive chemicals. Conventional cotton is developed in a profoundly lethal process—polluting groundwater and harming the evolved way of life [19] (Table 1).

10 Advantages of Organic Cotton Production

See Fig. 2.

11 Disadvantages of Organic Cotton

Organic cotton is more costly to create—outcome because of a six-year ponder in the USA demonstrated natural cotton generation costs at around half higher than

Table 1 Difference between organic and conventional cotton

Organic cotton	Conventional cotton
Genetically modified organisms are not involved in production process	Genetically modified organisms are actively involved
Organic manures, natural pesticides and insecticides are used	Chemical fertilizers and pesticides are used
Detoxification of soil and land occur invariably	Land detoxification does not occur
Multiple crop rotation can be practised	Only one crop culturing is done
Seed treatment is not necessary	Seed treatment with insecticides and pesticides is done prior culturing
Eco-friendly and does not cause any harm to living creatures	Results in adverse health issues due to over application of toxic chemicals
Protects biodiversity	Ecological balance is disturbed

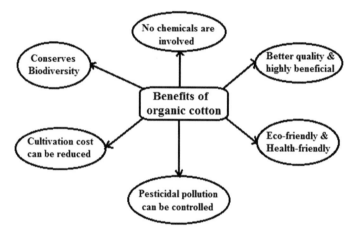

Fig. 2 Advantages of organic cotton

those of customary cotton. Organic cotton generally has bring down yields, which requires more land to create a similar amount of cotton and can have bring down evaluations, which influences financial aspects; and it requires altogether more work to deliver. For instance, the developing of organic cotton for the most part requires numerous specialists with tools to execute weeds, which expands work and generation costs. It likewise can require more vitality than traditional cotton creation due to increase in culturing. Contrasts in cotton creation systems ought to be considered while surveying the maintainability of organic cotton versus traditional cotton [20].

12   Limitations of Natural Cotton

There are numerous reasons why organic cotton generation has not stretched out to different nations. In any case, a significant number of them have officially ceased, not for absence of demand or interest for such cotton, however for economic reasons. Bug sprays should be wiped out from the cotton creation framework since they are unsafe to apply, have long haul results on the pest complex, and injurious consequences for the earth. The fundamental explanations for this confinement are:

- cost of generation of natural cotton versus traditional cotton, and
- cost premium on natural cotton.

The accompanying variables have restricted the extension of natural cotton creation. Reasonable measures must be received to advance production if natural cotton creation is to extend [21].

- **Appropriate Varieties**

Cotton manufacturers adjusted current assortments to organic generation practices. Monetarily developed assortments have been tried and produced for high input conditions. Assortments performing under ideal conditions will be unable to keep up their yield level without engineered composts and bug sprays. Reproducing material for natural cotton creation must be screened under natural conditions.

- **Utilization of Manures**

Assortments that are reasonable for high manure utilize have been developed under natural conditions. Therefore, such assortments have under gone heavier misfortunes in yield than anticipated, debilitating farmers from proceeding with natural creation. There is a need to create assortments appropriate for organic generation conditions, perhaps not as high yielding as would be expected assortments but rather strong and ready to deliver great yields under natural conditions. Assortments for natural creation must be produced under natural conditions. Engineered chemical manures are connected to cotton and to different products so as to address supplement issues for the plant. Supplement needs change from least to most extreme for N, P and K over the span of evolution. Nitrogen, which drains with water and can be lost through dissipation, must be connected when required for ideal plant development and organic product bearing. P and K can remain in the soil and be utilized when required, yet yields are genuinely influenced if the planning or measurement for nitrogen is changed. Green manuring and natural treatment can be utilized to keep up the required supplement supply, yet the accessibility of nitrogen to the level of inorganic preparation can't be accomplished.

- **Control of Pesticides**

The cotton plant is normally powerless against an assortment of insects, which attacks under natural developing conditions. Insect weight can be brought down by improving natural controls down to make up for the absence of bug spray utilize. The cotton plant has extraordinary systems compared to other worked in remuneration frameworks of numerous field crops. This happens in light of the fact that the plant turns out to be physiologically depleted and can't complete physiological procedures at the required rate, or due to changes in surrounding conditions and when natural growth of the plants is not permitted.

- **Generation Process and Cost**

Natural cotton agriculturists require counsel, without which they can be taking a chance with their venture. Sadly, data on cost of creation of organic cotton versus traditional generation under different arrangements of production conditions isn't accessible. Without such data, agriculturists are hesitant to change to natural organic cultivation.

- **Requirement for Alternate Inputs**

Engineered composts and bug sprays were received as a result of the high advantage cost proportions. The effect of compost and bug spray utilize is fast and extremely viable. The disposal of engineered composts and pesticides denies the plant two noteworthy shields, i.e. security against supplement starvation and defence against bugs, unless elective frameworks with similarly brisk and compelling activity are accessible. Shockingly such choices are not accessible. Manual and mechanical methods for weed control exist yet they are not achievable for vast scale cultivating frameworks, and other options to bug sprays and manures are moderate in real life [22].

13 Conclusion

There keeps on being overall enthusiasm for organic cotton as a conceivable eco-friendly approach to deliver cotton and for financial reasons. Generation of organic cotton has expanded as of late to around 0.1% of world cotton creation, mostly because of expanded generation in Turkey and in addition India, China and some African nations. Natural generation isn't really any more ecologically friendly or practical than current ordinary cotton creation. From a customer build-up outlook there is no distinction between traditionally developed cotton and naturally developed cotton. There are confinements to organic cotton generation that should be overcome if organic cotton is to wind up in excess of a little specialty advertise. Developing organic cotton is more requesting and more costly than developing cotton ordinarily which may be a drawback. Organic creation can be a genuine test if pest weights are high. With responsibility and experience, it might be conceivable and could give value premiums to cultivators willing to address the difficulties. Traditional and organic cotton generation can coincide. Profit percentage will drive choices in the supply chain network.

References

1. Wakelyn PJ, Chaudhry MR (2007) 5—organic cotton. In: Cotton, pp 130–175
2. Makita R (2012) Fair trade and organic initiatives confronted with Bt cotton in Andhra Pradesh, India: a paradox. Geoforum 43:1232–1241
3. Mohan KS, Ravi KC, Suresh PJ, Sumerford D, Head GP (2015) Field resistance to the *Bacillus thuringiensis* protein Cry1Ac expressed in Bollgard hybrid cotton in pink bollworm, *Pectinophora gossypiella* (Saunders), populations in India. Pest Manag Sci 72(4):738–746
4. Vigneshwaran D, Ananthasubramanian M, Kandhavadivu P (2014) 7—bioprocessing of organic cotton textiles. In: Bioprocessing of textiles, pp 319–397
5. Fox sv. (1994) Organic cotton. In: Proceedings of the international cotton conference, Bremen, pp 317–319
6. Radhakrishnan S (2017) 2—sustainable cotton production. Sustain Fibre Text, 21–67
7. Soltani S, Azadi H, Mahmoudi H, Witlox F (2014) Organic agriculture in Iran: farmers' barriers to and factors influencing adoption. Renew Agric Food Syst 29:126–134
8. Hae Now (2006) Organic cotton clothing 'Why choose organic', http://www.haenow.com/wh yorganic.php
9. Sharma SK, Bugalya K (2014) Competitiveness of Indian agriculture sector: a case study of cotton crop. Proc Soc Behav Sci 133:320–335
10. Wakelyn PJ, Chaudhry MR (2009) 11—organic cotton: production practices and post-harvest considerations. In: Sustainable textiles, pp 231–301
11. Marquardt S (2003) Organic cotton: production and market trends in the United States and Canada—2001 and 2003. In: Proceedings of the 2003 beltwide cotton conference. National Cotton Council, Memphis, TN, pp 362–366
12. Myers D, Stolton S (eds) (1999) Organic cotton—from field to final product. Intermediate Technology Publications, Intermediate Technology Development Group, 103/105 Southampton Row, London, WC1B 4HH, UK
13. Demeritt L (2006) Behind the buzz: what consumers think of organic labelling. Org Process 3(1):14–17
14. Adanacioglu H, Olgun A (2012) Evaluation of the efficiency of organic cotton farmers: a case study from turkey. Bul J Agric Sci 18:418–428
15. Altenbuchner C, Larcher M, Vogel S (2014) The impact of organic cotton cultivation on the livelihood of smallholder farmers in Meatu district, Tanzania. Renew Agric Food Syst, pp 1–15
16. Lakerveld RP, Lele S, Crane TA, Fortuin KPJ, Springate Baginski O (2015) The social distribution of provisioning forest ecosystem services: evidence and insights from Odisha, India. Ecosyst Serv 14:56–66
17. Cotton Inc. (2012) Life cycle assessment of cotton fibre and fabric. Prepared for VISION 21, a project of the cotton foundation and managed by Cotton Incorporated, Cotton Council International and The National Cotton Council. The research was conducted by Cotton Incorporated and PE International, http://cottontoday.cottoninc.com/Life-Cycle-Assessment/
18. Branson A, Sood D (2015) India—cotton and products annual 2015. USDA Foreign Agricultural Service—GAIN Global Agricultural Information Network, Washington, DC
19. Patil S, Reidsma P, Shah P, Purushothaman S, Wolf J (2014) Comparing conventional and organic agriculture in Karnataka, India: where and when can organic farming be sustainable? Land Use Policy 37:40–51
20. Bello WB (2008) Problems and prospect of organic farming in developing countries. Ethiop J Environ Stud Manage 1:36–43
21. Chaudhry MR (2003) Limitations on organic cotton production. In: The ICAC recorder, vol XXI, p 1
22. Bachmann F (2012) Potential and limitations of organic and fair trade cotton for improving livelihoods of smallholders: evidence from Central Asia. Renew Agric Food Syst 27(02):138–147

Organic Cotton Versus Recycled Cotton Versus Sustainable Cotton

P. Senthil Kumar and A. Saravanan

Abstract Organic cotton will be cotton that is relied upon to have been developed without manures and pesticides, with rehearses that advance biodiversity, organic cycles, and soil health. As far as natural cotton, China, Turkey, and India are the world's driving makers and sources. While natural cotton makes cotton development "cleaner," both natural and ordinary cotton experience a similar assembling process, which is water and vitality concentrated. Recycled cotton is re-purposed, post-modern or post-shopper cotton that would somehow or another is considered straight up: squander for the landfill. The pieces of such cut and sew offices are post-mechanical cotton "squander" that have the ability to be reused. Contingent upon how reused cotton is utilized, it can possibly extraordinarily decrease water and vitality utilization in reasonable design and attire, and diminish landfill waste and space. Cotton development is related with various social, financial and natural difficulties that debilitate the part's sustainability. Development of more reasonable cotton has never been higher than it is today, achieving 2.6 million tons in 2015/16, around 12% of aggregate worldwide supply. Nonetheless, just barely finished a fifth (21%) of this sum is effectively sourced as more sustainable cotton by organizations with the rest of as customary cotton.

Keywords Organic cotton · Biodiversity · Landfill · Recycling · Sustainable

1 Introduction

Cotton is a fibre that we tend to utilize a considerable measure of. It develops normally, inhales well and structures the spine behind a considerable measure of our staple articles of clothing—a valid example: white T-shirts and denim pants. In any case,

P. Senthil Kumar (✉)
Department of Chemical Engineering, SSN College of Engineering, Chennai 603110, India
e-mail: senthilkumarp@ssn.edu.in

A. Saravanan
Department of Biotechnology, Rajalakshmi Engineering College, Chennai 602105, India

© Springer Nature Singapore Pte Ltd. 2019   141
M. A. Gardetti and S. S. Muthu (eds.), *Organic Cotton*, Textile Science
and Clothing Technology, https://doi.org/10.1007/978-981-10-8782-0_7

not all cotton was made equivalent. Truth be told, there are entirely of various kinds of cotton and these can be depicted in view of how they were delivered. Every one of these creation approaches has fluctuating natural and social effects, which is the reason to trust it is essential to comprehend the contrasts between conventional, organic and recycled cotton [14].

Cotton cultivation is related with various social, monetary and ecological difficulties that debilitate the area's supportability. Development of more practical cotton has never been higher than it is today, achieving 2.6 million tons in 2015/16, around 12% of aggregate worldwide supply. In any case, just barely finished a fifth (21%) of this sum is effectively sourced as more supportable cotton by organizations with the rest of as traditional cotton. Current gauges are for more reasonable cotton to be no less than 15% of worldwide cotton creation in 2017.

Organizations that depend to a great extent on cotton as a crude material assume an essential part in securing the eventual fate of the manageable cotton showcase, decreasing cotton's natural effects and enhancing work conditions. Some have made open time bound duties regarding more sustainable cotton sourcing and report unfaltering advancement in genuine take-up of reasonable cotton. However a significant number of the world's biggest organizations utilizing cotton as a key crude material don't consider or address the negative effects of its creation.

1.1 A Key Change Is Important to Take Care of Regular Worldwide Issues

Normally, 99% of organizations creating retail items are in reality endeavouring to tackle an issue that can't be fathomed, unless they will make strides and survey what they do, need to be, and where to go later on. A worldwide natural issue can't be comprehended by little repairs and conceal of breaks. The issue can't be explained when the significant proprietors, financial specialists continues requesting a bigger return of their speculations consistently. On the off chance that you read in the diverse real brands organization's main goal, mark esteems, investor data and business methodologies the following 2–5 years you will find that they make them thing in like manner, extending and development of business.

1.2 Sustainability Challenges

Ordinary cotton development is portrayed by interconnected ecological, social and financial challenges that debilitate the segment's general sustainability. Abuse of manufactured composts likewise causes loss of soil richness and soil fermentation. Weight ashore and arrive freedom because ecological effects, for example, soil disintegration, soil defilement, and biodiversity misfortune. An absence of sexual orientation value, a typical issue in the agrarian division, is keeping down comprehensive improvement in cotton cultivating groups [9, 12]. Worldwide exchange structures are by and large ominous for ranchers. What's more, numerous ranchers are in the red because of the overwhelming use of exorbitant data sources. Every one of these components add to propagating destitution for numerous cotton agriculturists. In spite of these difficulties, cotton still furnishes ranchers with a money salary that pays for building material, school expenses, and other family unit necessities. On the off chance that created economically, cotton can give a huge number of cotton ranchers overall additional salary also, enables them to enhance their lives.

1.3 Sustainable Options

Supportability benchmarks and projects intend to address the challenges related with regular cotton development. They give direction to agriculturists on more supportable cultivating hones and guarantee purchasers that the item meets determined prerequisites. In spite of the fact that not by any means the only method to increment sustainability, such norms give a prompt initial step that all organizations can take. While sustainability can be sought after outside the system of a standard, sustainable development norms are a effective device interfacing economical cultivating hones, advertise request and claims [3, 10]. As this report centres around showcase request and the part of organizations in driving interest, solid economical cotton development guidelines are focal to its examination. They are viewed as dependable both in terms of their substance and the frameworks that manage how they are actualized, evaluated and represented.

While a few organizations endeavour to set the right case, most seem to do pretty much nothing or nothing to address the sustainability issues related with cotton developing. There is along these lines huge opportunity to get better in organization sourcing and giving an account of sustainable cotton.

1.4 Announcing

This report features positive advancements and results accomplished by a few organizations, in any case, unmistakably exhibits the across the board nonattendance of openly accessible data on reasonable cotton sourcing.

(i) **Strategy**

There is a critical absence of data on sustainable cotton arrangements. Norms vital part in tending to water utilize, biodiversity, dangerous pesticides and work rights issues. Be that as it may, few organizations determine what cotton-related approach measures they actualize past this. Work rights and reusing get less consideration than natural issues in organizations' manageable cotton arrangements. At last, just a couple of organizations indicate clear and timebound focuses for more reasonable cotton sourcing. Altogether, the organizations with a unmistakably characterized target scored best in the generally evaluation.

(ii) **Traceability**

Open data with respect to non-guaranteed cotton and the geographic starting points of cotton is uncommon. Data is additionally constrained concerning chain relations at the last generation arrange, and particularly the texture and yarn producing stage.

(iii) **Other Consideration**

Conversely, even high scoring organizations offer restricted data with respect to traceability for their whole cotton supply, inventory network relations, total volumes of cotton sourced and nations of cause. While there are various organizations that work difficult to set the correct case, numerous driving organizations and their brands in the cotton business still have much space for natural and social change. The majority of the organizations examined don't have clear arrangements in regards to a more supportable cotton supply. These organizations don't show up to organize supportability or moral business with respect to supportability of their cotton supplies, or if nothing else neglect to enough report on their endeavours to do as such.

1.5 *Social and Exchange Related Parts of Sustainable Cotton Creation and Preparing*

This part acquaints key viewpoints with social supportability—work conditions, word related wellbeing and security, human rights and to a specific degree, exchange issues. Social issues are an critical part of supportability for ideological reasons, yet in addition for simply business or advertising closes. For instance, shoppers would likely be horrified if they somehow happened to learn that the natural or "economical" cotton shirt or pants, for which they had paid a premium, were delivered in sweat shop conditions or utilizing youngster work. All things considered, there are down to earth also as moral purposes behind tending to social equity along the production network. It can be convincingly contended that all together for the cotton store network to be genuinely reasonable it must address social and financial issues, and in addition ecological.

1.6 *Potential Socially and Earth Sustainable Cotton Supply Chains*

Cotton is a perplexing item from various perspectives. It joins social and ecological concerns at numerous focuses along the store network and will request imaginative facilitated endeavors keeping in mind the end goal to genuinely guarantee and final result that is sustainable in both social and natural domains [2]. The pilot review encounter canvassed the accompanying strides in the chain: creation (e.g. agriculturist fields and helpful administration); ginnery; turning; passing on; and cut-make-trim. The review tried to think about how the social and ecological accreditation frameworks could facilitate keeping in mind the end goal to give a completely affirmed final result for buyers. A portion of the diverse strides in the preparing/store network and accreditation alternatives are given as cases of how each level may be guaranteed so as to guarantee that social and natural criteria are met. This is in no way, shape or form an authoritative guide but instead a device to fortify additionally thought on coordination for a more practical cotton end item.

1.7   Conventional Cotton

Numerous individuals consider cotton a statement unquote 'regular' fibre, however when developed customarily, this apparently safe harvest can really wreak a great deal of destruction. Taking up a genuinely little level of the world's territory these yields represent an excessively huge level of the world's harmful chemicals [18]. That rate has really begun to decrease in a few sections of the world, yet as indicated by the Cotton Advisory Committee (CAC), cotton still speaks to as much as 5% of all pesticides and 14% of all bug sprays utilize all inclusive. Likewise, it takes around 290 gallons of water to develop enough ordinary cotton to deliver one T-shirt. So not exclusively does the majority of this negatively affect the earth, at that point; it can likewise be annihilating for ladies' wellbeing as well [17]. Figure 1 show the life cycle of conventional cotton yarn.

1.8   The Life Cycle of Conventional Cotton Yarn

- **The development organizes**: Concoction items expended in the life-cycle of customary cotton versus reused cotton in the development organize
- **The coloring stage**: Concoction items expended in the life-cycle of regular cotton versus reused cotton in the coloring stage
- **Discharge**: Outflow to the environment in life cycle of traditional cotton versus reused cotton
- **Waste**: Squander in the life-cycle of customary cotton yarns versus reused cotton yarn. Measure of water devoured in the life-cycle of regular cotton versus reused cotton yarn in the coloring procedure
- **Wastewater**: Water squander gushing in the life-cycle of regular cotton versus reused cotton yarn in the coloring procedure
- **Vitality utilization add up to thought about**: Vitality utilization in the life-cycle of traditional cotton yarn versus reused cotton yarn.

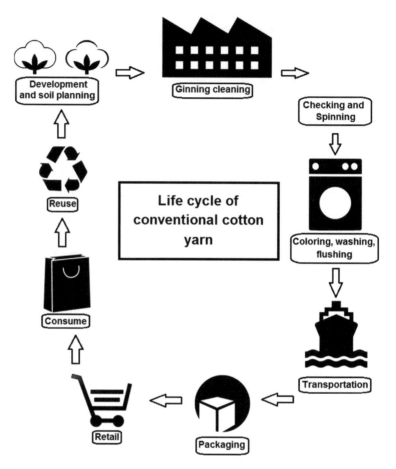

Fig. 1 Life cycle conventional cotton yarn

1.9 Recycled Cotton

Recycled cotton is re-purposed, consider straight up post-modern or post-customer cotton that would some way or another: waste for the landfill. For instance, envision an industrial facility that removes designs for dress of enormous sheets of texture and has all the in the middle of pieces to dispose of. The pieces of such cut and sew offices are post-mechanical cotton "Waste" that have the ability to be reused [13]. Contingent upon how reused cotton is utilized, it can possibly enormously decrease water and vitality utilization in feasible form and clothing, and in addition diminish landfill waste and space [7].

Recycled yarn or fabric of cotton, wool, acryl or viscose is limited available for apparel manufacturing. Some mixed fabrics (blends of recycled and virgin) can be available for textile recycling. Attempting to re-reason fabricated cotton that would some way or another be discarded, Recover utilizes reused cotton in their 100% reused economical attire. The manner by which Recover utilizes post-mechanical reused cotton is better portrayed as "upcycling." Rather than debasing nature of item in the reusing procedure, upcycling includes taking undesirable material that would be waste and improving it into an item [4, 5]. By utilizing reused cotton in Recover items and improving generation, Recover's assembling procedure brings about a 35% diminishment in ozone harming substance outflows, 66% lessening in vitality utilization, and 55% decrease in water utilization contrasted with that of a traditionally colored shirt. Continually taking a gander at the 10,000 foot view, Recover keeps plastic containers and post-mechanical cotton scraps from setting off to the landfill and re-purposes them to make agreeable, delicate, solid, and sleek attire. Figure 2 show the high quality recycled cotton.

Shockingly, reusing cotton garments isn't straightforward. To make another bit of dress from old garments, the old garments initially must be slashed up and transformed over into crude material. However, that slashing up process tends to bring down the cotton's quality since it abbreviates the staple length of the strands. Staple length assumes a critical part in deciding the quality and non-abrasiveness of cotton strings. The more drawn out the staple, the better these qualities, which are the reason cotton assortments with additional long staple lengths. In any case, obviously figuring out how to reuse cotton effectively, without weakening it, could be the kind of "distinct advantage" the establishment is looking for, and an opposition could accelerate the procedure. "The biggest potential lies with finding new innovation that implies we can reuse strands with unaltered quality".

There are really two distinct sorts of recycled cotton—post-customer and pre-purchaser reused cotton. The previous is reutilised after the shopper is done with their article of clothing, while the last is produced using producer squander. This implies the pieces, rejects and trimmings, which for the most part wouldn't make it into the client's hands by any stretch of the imagination—however would typically be disposed of—can be given new life. The two types of recycled cotton speak to a superior option than essentially sending texture waste to landfill, yet pre-shopper cotton compensates for the way that an expected 15–20% of texture is squandered all through the creation procedure.

Reused cotton is another practical alternative. Cotton can be reused from pre-shopper squander created amid the material creation process and from post-customer squander including disposed of material items. Cotton is as of now reused generally through mechanical fibre reusing which minimize fibre length and quality and requires mixing with virgin filaments for additionally utilize. Rising advancements offer some expectation for substance cotton fibre reusing that would permit creation of reused fibre break even with in quality to virgin fibre. Figure 3 show the life cycle conventional cotton yarn.

Two fundamental gauges created by Textiles Exchange give systems for exchanging and preparing reused cotton:

Fig. 2 High quality recycled
cotton

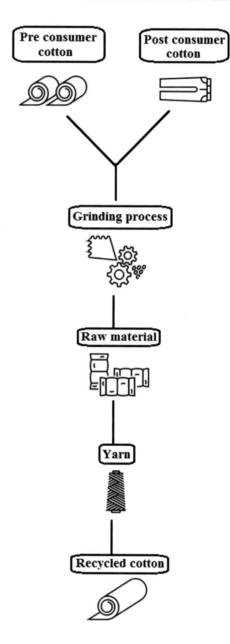

- Global Recycle Standard (GRS)—a comprehensive confirmation standard for items containing any reused content;
- Reused Claim Standard (RCS)—a chain of guardianship standard to track reused crude materials through supply chains.

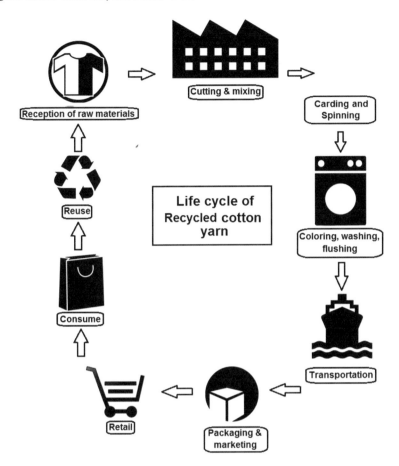

Fig. 3 Life cycle of recycled cotton yarn

1.10 Recycled Cotton Yarn Advantage

Recycled cotton is a great alternative to regular cotton, since it uses no water or chemicals in its manufacturing process and requires no new farmland. Good quality clothing is sent to charity institutions and is used as second hand clothing. Unwearable textiles are considered as damaged textiles, and are processed in the factory as rags. Rags are collected and sent to the wiping and flocking industry. Other materials will be sent for fibre reclamation and stuffing. Fibres from the old fabrics are reclaimed and are used for making new garments. Threads from the fabric is pulled out and used for re-weaving new garments or blankets. Both natural and synthetic fibres can be recycled this way. Incoming textiles are graded into type and color. Initially the material is shredded into fibres called shoddy. Later based on the end use, other fibres are blended with shoddy. The blended mixture is carded, and spun for weaving or knitting.

1.11 Recycled Cotton Is the Most Environmentally Friendly Option

In research and concentrates done, the utilization of reused cotton yarn has the upsides of devouring less water and concoction items. There are fewer stages in the generation life-cycles; it doesn't defile the subsoil, water, or air. Most of all the reused textures or article of clothing would somehow or another end up squander, gives financial advantages to creating nations, which been genuinely harmed as of late on account of exorbitant large scale manufacturing of cotton [11]. The general expend of vitality is the most minimal and the truth of the matter is that the generation costs are lower than when utilizing ordinary or completely affirmed natural cotton. Recycled cotton is a product that comes from unnecessary waste [1, 8]. This can be recycled from pre consumer cotton waste, the excess material produced during the production of yarn or from post consumer cotton that comes from discarded textile products [6].

1.12 Process of Recycling

- Approaching material is evaluated into sort and shading (mind is taken to guarantee the wellspring of the approaching material is protected from contaminants)
- The materials are destroyed and mixed with other chose filaments of comparative shading, contingent upon the expected utilization of the reused yarn. Regularly reused cotton yarns are mixed in a blend comprising of ~20% polyester yarn, to give extra quality to the subsequent texture
- The mixed blend is cleaned and faded if vital.

1.13 The Challenge

Despite the fact that numerous have attempted, few have prevailing at reusing cotton productively. Before old material can be transformed into something new, it must be ground into modest bits and restored as crude material. Sadly, this procedure is famous for making a result of low quality. Have you encountered crude, reused cotton? It is normally delicate, unpleasant, not happy or sufficiently tough for rehashed wear and it's in this manner hard to utilize a huge amount of reused cotton in a solitary piece of clothing. This implies it for the most part should be blended with different kinds of strands.

1.14 Organic Cotton

Organic cotton will be cotton that is relied upon to have been developed without manures and pesticides, with hones that advance biodiversity, organic cycles, and soil wellbeing. Cotton overall is thought to be one of the world's "dirtiest" harvests, covering 2.5% of the world's developed land and representing 24% ($2.6 billion worth) of the world's bug spray showcase. As far as natural cotton, China, Turkey, and India are the world's driving makers and sources [16]. While natural cotton makes cotton development "cleaner," both natural and regular cotton experience a similar assembling process, which is water and vitality concentrated.

Organic cotton is developed utilizing techniques and materials that have a low ecological effect. Intended to diminish the utilization of harmful pesticides and composts normally utilized as a part of regular cotton creation, this approach additionally renews soil richness and look after biodiversity, instead of stripping the earth of these characteristics [15]. There are strict benchmarks controlling natural cotton order, yet while it is surely more supportable than regular cotton, regardless it does not speak to an impeccable option. Natural cotton harvests can here and there be considerably more asset serious than their customary partners; requiring as much as 660 gallons of water with a specific end goal to create one T-shirt. Therefore, it is yet essential to purchase just the same number of traditional cotton pieces as you truly require.

Organic horticulture (nourishment and fiber) secures the wellbeing of individuals and the planet by decreasing the general presentation to lethal chemicals from engineered pesticides that can wind up in the ground, air, water and sustenance supply, and that are related with wellbeing outcomes, from asthma to malignancy. Since natural farming doesn't utilize dangerous pesticides, picking natural items is a simple method to help ensure the earth and yourself.

Organic cotton is grown using methods and materials that have a low impact on the environment. Organic production systems replenish and maintain soil fertility, reduce the use of toxic and persistent pesticides and fertilizers. Organic cotton is the precursor of sustainable items. It is produced in sustainable systems to maintain natural resources, without the use of any chemicals. The main difference from traditional cotton resides in the enormous environmental benefits. The distinction is a blend of numerous variables above all the diverse life cycle between customary cotton yarns and recyclable cotton yarns.

There has been developing enthusiasm for organic cotton in the worldwide clothing industry in the previous couple of years, with defenders asserting an assortment of ecological and social advantages over the esteem chain. These cases were regularly met with suspicion from different quarters that were agreeable to the norm. Nevertheless, an exhaustive report by the textile exchange covering the main five cotton developing countries has uncovered the accompanying discoveries of the effect of natural cotton development over customary cotton:

- 46% reduced a worldwide temperature alteration potential
- 70% diminished fermentation potential
- 26% diminished eutrophication potential (soil disintegration)

- 91% diminished blue water utilization
- 62% diminished essential vitality request (non-inexhaustible).

The accentuation today is on naturally well disposed creation hones also, 'feasible' creation through the whole material chain. Consider the following: is organic cotton more 'economical' than regular cotton; an naturally best item, of added benefit to the earth, agriculturists, and shoppers; or is it basically a showcasing apparatus or is it both? Defenders of natural cotton and the individuals who showcase natural cotton items inaccurately advance the observation that regular cotton isn't an ecologically dependably delivered crop.

1.15 Organic Cotton Is Getting Acknowledgment

Accordingly the utilization of organic cotton that has no utilization of chemicals in the procedures of making is to be favoured. The business sectors have slowly acknowledged to utilize organic cotton and it's triumphant both homestead and buyers. Notwithstanding, a glance at the extent of the dress market all around it is that to change this over from unadulterated regular cotton to organic cotton is mission unfeasible.

1.16 Employments of Organic Cotton

Organic cotton can properly be known as the most skin-accommodating, most calming, and most safe regular fiber. While traditional cotton can once in a while be bothering to infant skins, Organic Cotton is never similar to that. It is the perfect material for securing and cleaning infants, especially to make garments, gauzes, covering and cleaning wounds, bassinet sheets, infant garments, towels, and a great many such things. It can likewise be securely utilized as a part of surgeries where sullying from any source can be lethal. Natural Cotton Seed Oil, a side-effect of Organic Cotton, has wide uses in snacks and in sustain for domesticated animals.

1.17 Organic Cotton—Benefits

Organic cotton demonstrates extraordinary advantages at different levels of the esteem chain. Figure 4 show the benefits of organic cotton.

Ranchers, merchants, retailers and purchasers all advantage from the financial, social and natural points of interest of natural cotton ventures. Tap on the structures to see the advantages for each gathering.

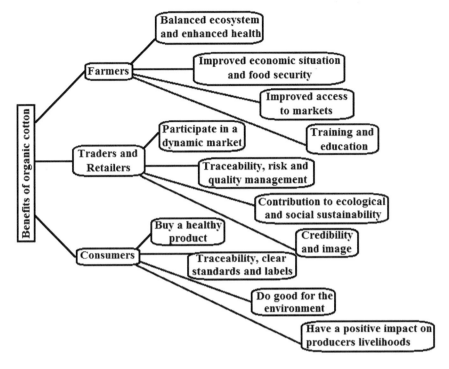

Fig. 4 Benefits of organic cotton

1.18 Advantages

(i) Eco Print

Organic cottons are developed utilizing techniques and materials that lowly affect the earth. Natural generation frameworks recharge and keep up soil richness, lessen the utilization of harmful and determined pesticides and manures, and assemble organically differing farming, making normal cottons an eco well disposed texture.

(ii) Dampness Absorbent

Natural cottons are valuable in material creation because of its characteristic wicking properties, assimilation of colour shading and its capacity to settle other eco filaments.

(iii) Feel of Fabric

Garments produced using natural cottons have the vibe of material without the weight. Since organic cottons are not synthetically stripped of its normal wax, most weaves have a trademark smoothness and weight which makes the texture especially complimenting in its wrap and in the smooth way it reflects and retains light.

(iv) **Anti Allergenic**

Numerous sensitivities might be caused by chemicals in cotton strands we either wear, or consider.

1.19 Comparison of Use of Water for a Conventional Cotton T-Shirt and Organic Cotton T-Shirt

Conventional cotton: Water use per pound of cotton: 0.002 AF.
Organic cotton: Water use per pound of cotton: 0.0024 AF.

Organic cotton is a rotation crop. When crops are rotated, the soil maintains its nutrients and is better able to hold water in. Regular cotton is usually the sole crop planted. Cotton depletes the soil, and leaves the soil incapable of holding water.

Most organic cotton is rain-fed and not irrigated. Crop rotation is also an effective measure to break many insect pests and plant disease cycles. Another plus is that organic cotton farms keep many people employed fairly. The water use of organic cotton compared to conventional cotton is indeed a slightly higher (0.0004 AF). But after 1 or 2 rotation cycles the soil quality rises and allows for the same or even less water usage. As organic cotton can be rain-fed in a lot of regions in the world energy use in the production process is also reduced. Other advantages like the use of biological systems to keep the balance, instead of using synthetic agricultural chemicals are obvious.

1.20 Conclusions and Recommendations

The material and attire industry is advancing toward a roundabout economy and shut circle producing. To accomplish this, one possible strategy is the utilization of reused materials in material and clothing items. The fate of reusing depends intensely on the improvement of new propelled innovations and methodologies for material handling (without quality misfortune), accumulation, arranging, preparing, and use in another item that is additionally recyclable. Making an interest for new items with reused content is basic. It is imperative to incorporate the reused content in the plan and item advancement phases of design and home items in any case; there is additionally a need to energize streams that advance reusing and reuse.

Cotton is usually classified as plain, twill and satin weaves according to the way it is woven. Organizations that utilization vast volumes of cotton in their items have a key part to play, in empowering the further development of the reasonable cotton market, and supporting ranchers to change to more sustainable types of development. To satisfy this part, organizations should begin or proceed sourcing more feasible cotton and, as a need:

- Receive strategies on general cotton supportability, and particularly on the key themes of water, biodiversity, work conditions, and reusing;
- Set open focuses for utilizing 100% practical cotton by 2020, including the level of Better Cotton, natural, Fair trade and reused cotton;
- Report straightforwardly every year on approaches, methodologies and focuses, and in addition execution and advance.

In help of the above, organizations ought to likewise:

- Cotton supply ties to at any rate the nation of development;
- Figure the volume of cotton utilized by the business yearly and make it open;
- Build up a far reaching get ready for applying arrangements and meeting sourcing targets, including every single significant office.

References

1. Babu KM, Selvadass M (2013) Life cycle assessment for the dyeing and finishing process of organic cotton knitted fabrics. J Text Apparel Technol Manage 8:1–16
2. Bevilacqua M, Ciarapica FM, Mazzuto G, Paciarotti C (2014) Environmental analysis of a cotton yarn supply chain. J Clean Prod 82:154–165
3. Chen L, Wang B, Ruan X, Chen J, Yang Y (2015) Hydrolysis-free and fully recyclable reactive dyeing of cotton in green, non-nucleophilic solvents for a sustainable textile industry. J Clean Prod 107:550–556
4. Collier BJ, Collier JR, Petrovan S, Negulescu II (2007) Recycling of cotton. Cotton 15:484–500
5. Durham E, Hewitt A, Bell R, Russell S (2015) Technical design for recycling of clothing. Sustain Apparel 7:187–198
6. Esteve-Turrillas FA, Guardia MDL (2017) Environmental impact of recover cotton in textile industry. Resour Conserv Recycl 116:107–115
7. Halimi MT, Hassen MB, Sakli F (2008) Cotton waste recycling: quantitative and qualitative assessment. Resour Conserv Recycl 52(5):785–791
8. Jackson GJ (2005) Organic cotton farming in Kutch. Agrocel, Gujarat, India
9. Khor LY, Feike T (2017) Economic sustainability of irrigation practices in arid cotton production. Water Resour Econ 20:40–52
10. Krishna VV, Qaim M (2012) Bt cotton and sustainability of pesticide reductions in India. Agric Syst 107:47–55
11. Leifeld F (1996) Cotton waste reclamation. Recycl Text Plast Waste, 107–119
12. Radhakrishnan S (2017) Sustainable cotton production. Sustain Fibres Text 2:21–67
13. Rajput D, Bhagade SS, Raut SP, Ralegaonkar RV, Mandavgane SA (2012) Reuse of cotton and recycle paper mill waste as building material. Construct Build Mater 34:470–475
14. Rana S, Karunamoorthy S, Parveen S, Fangueiro R (2015) Life cycle assessment of cotton textiles and clothing. Handb Life Cycle Assess (LCA) Text Clothing 9:195–216
15. Vigneswaran C, Ananthasubramanian M, Kandhavadivu P (2014) Bioprocessing of organic cotton textiles. Bioprocess Text 7:319–397
16. Wakelyn PJ, Chaudhry MR (2007) Organic cotton. Cotton 5:130–175
17. Wossink A, Denaux ZS (2006) Environmental and cost efficiency of pesticide use in transgenic and conventional cotton production. Agric Syst 90(1–3):312–328
18. Zulfiqar F, Thapa GB (2016) Is 'better cotton' better than conventional cotton in terms of input use efficiency and financial performance? Land Use Pol 52:136–143

Organic Cotton and Environmental Impacts

Seyda Eyupoglu

Abstract Recently, environmental consciousness and demand of healthy food have increased rapidly. In this context, organic agriculture has been used in all countries. In addition, production area and producers have increased each passing days. "Organic agriculture" also known as "ecological agriculture or biological agriculture" is a production method that aims eco-friendly production, development of plant resistance, and conservations of agriculture soil. Besides, organic agriculture aims to reconstruct of lost natural balance. Organic agriculture bans use of pesticides, hormones, and chemical fertilizers. Especially, organic agriculture gained a commercial dimension with the increasing of consumer demands in 1980s. In organic cotton agriculture, genetically modified cotton seeds are not used. The fundamental of organic cotton agriculture is that cotton seeds are not treated microwave energy and radiation. Organic cotton agriculture consists of all agriculture systems that encourage eco-friendly fiber production. Furthermore, it causes to remove use of chemical fertilizers, pesticides and pharmaceuticals. For this reason, organic agriculture causes to increase the fertility of soil. In this chapter, organic agriculture, organic cotton agriculture, comparison conventional cotton agriculture with organic cotton agriculture, environmental impacts of organic cotton agriculture, and use of organic cotton products were investigated.

Keywords Organic cotton · Organic agriculture · Environmental impacts

1 Introduction

Cotton plants belong to the genus *Gossypium* in the mallow family *Malvaceae*. Cotton plants are a broad-leaves plants and each of seeds are covered with white or cream color hairs. These hairs can be classified as short and long hairs. Short hairs are

S. Eyupoglu (✉)
Department of Fashion and Textile Design, Faculty of Architecture and Design, Istanbul Commerce University, Istanbul, Turkey
e-mail: scanbolat@ticaret.edu.tr

© Springer Nature Singapore Pte Ltd. 2019
M. A. Gardetti and S. S. Muthu (eds.), *Organic Cotton*, Textile Science and Clothing Technology, https://doi.org/10.1007/978-981-10-8782-0_8

Table 1 The data of cotton in the world

	2012–2013	2013–2014	2014–2015	2015–2016	2016–2017[a]	Variation (%)
Area (thousand hectare)	34.149	32.647	33.929	30.490	30.085	−1.3
Yield (ton/hectare)	784	802	772	693	745	7.5
Production	26.785	26.175	26.201	21.030	22.480	6.9
Consumption	23.780	24.004	24.465	24.170	24.200	0.1
Ending stocks	18.508	20.604	22.315	19.100	17.380	−0.9
Imports	10.201	8.935	7.785	7.460	7.640	2.4
Exports	10.061	9.010	7.808	7.530	7.640	1.5

[a]Estimation

called as linter and long hairs are called as fiber. Cotton plants like warm and moist climate. Furthermore, for optimum development, these plants need warm climatic conditions, enough sunlight and plenty of moisture for 6–7 months. Cotton plants come into flower 80–90 days after seeds planting. Fibers are cultivated as soon as possible in order to prevent deterioration quality of fibers with light and moisture. The right time of the cotton harvest affects the quality of the fibers [1].

Cotton plants have been used during ancient times for fibers. Cotton fibers have been spun, weaved and dyed. In India, cotton fabric and yarn samples which belong to B.C. 3000 were made out an excavation. Furthermore, the remains of cotton plants were found in Egypt. Indian was head quarts of cotton industry in B.C. 1500 to 16th century and cotton fabrics were dealt in Mediterranean Countries. Cotton fibers were begun to use by Greeks after India was invaded by Alexander the Great. In Europe, Spain and Italy are the first countries that cultivated cotton plants and produced of cotton goods. Arabian merchants brought cotton fabrics such as muslin and calico. Cotton fibers were introduced to China and Japan by Indians but widespread use of cotton fibers was quite slow. Japan began to cultivate of cotton plants in 17th century. Furthermore, England started to cultivate cotton plants in 1635. Native Americans used cotton fibers proficient and produced clothes fabrics. In Peru, cotton fabrics were found and it was estimated that these fabrics are belongs to a civilization before the Incas. Cotton fibers were used by American colonies such as Virginia, Carolina, and Georgia in 17th century. In 1793, cotton gin machine was invented and the revolution in the development of cotton has come to fruition [1].

At the present time, major cotton producers are China, United States of America and Indian. Pakistan, Brazil, Turkey, Uzbekistan, Egypt, Mexico, Iran and Sudan follow to these countries. The data of world cotton was given in Table 1 [2]. Furthermore, cotton cultivation area in the world was given in Table 2 [3].

In the world, the consumption of cotton is affected from economic growth, increase in population, rules of commerce, and fiber prices. Moreover, the consumption of cotton fiber can be affected average revenue of textile fiber consumption, petrol and

Table 2 Cotton cultivation area in the world

Country	2013/2014	2014/2015	2015/2016	2016/2017	2017/2018
India	11.650	12.846	11.638	10.845	12.235
USA	3.053	3.783	3.291	3.848	4.616
China	4.700	4.310	3.793	2.923	3.157
Pakistan	2.914	2.958	2.670	2.496	3.097
Uzbekistan	1.275	1.298	1.272	1.250	1.208
Brazil	1.010	976	1.007	939	1.555
Burkina Faso	644	661	631	740	770
Turkmenistan	545	545	534	545	534
Turkey	451	460	440	420	462
Argentina	506	456	447	247	305
Other	5.934	5.619	5.440	5.418	5.069

polyester prices. According to change by years, China, Turkey and Bangladesh are the largest cotton importers in the world. Tables 3 and 4 show the world cotton consumption and world cotton imports [4].

Cotton plant is an important plant that meets the basic raw material of many industries. It is raw material of textile industry under favour of fibers, raw material of feed industry under favour of pulp and oil produced from cotton seeds is raw material of vegetable oil industry. Figure 1 shows the usage areas of cotton plants.

Recently, protection of environment, human and community health consciousness have increased gradually. Environmental pollution which is classified as air pollution, sound pollution, industrial and nuclear wastes rises rapidly. In agriculture, the intensive use of chemical fertilizers and pesticides causes to damage the natural balance and it is estimated deadly hazards that food chain can reach with all livings. Because of increase in environmental pollution rapidly, organic agriculture which is eco-friendly agriculture system has become the main topic of conversation. Organic agriculture is described that it is not used synthetic insecticides, herbicides, fungicides and hormones. Organic agriculture which protects environment resists to organic fertilization and biological diseases. The aim of this system is high quality and efficiency.

Organic cotton is generally identified non-genetically modified cotton plant which is cultivated in subtropical countries as China, Turkey and United States, and without the use of any fertilizers and pesticides. Organic cotton is grown with eco-friendly methods and materials. Organic agriculture systems lead to freshen and continue soil fertility, block the use of toxic pesticides and synthetic fertilizers, and build biologically diverse agriculture. Organic cotton production is assumed to improve biodiversity and biological cycles. Furthermore, federal regulations avoid the use of genetically engineered seeds for organic farming.

Cotton covers 2.5% of the cultivated land of the world and the use of insecticides in cotton is 16% of the world insecticides. Compared with conventional cotton agri-

Table 3 Cotton consumption in the world

Word cotton consumption

Million metric tons	2013/14	2014/15	2015/16	2016/17	2017/18 February	2017/18 March
China	7.5	7.4	7.6	8.2	7.7	8.7
India	5.1	5.3	5.4	5.2	5.2	5.3
Pakistan	2.3	2.3	2.2	2.2	2.2	2.3
Bangladesh	1.2	1.3	1.4	1.5	1.4	1.6
Turkey	1.4	1.4	1.5	1.4	1.4	1.5
Vietnam	0.7	0.9	1.0	1.2	1.0	1.4
Indonesia	0.7	0.7	0.7	0.7	0.7	0.7
Brazil	0.9	0.7	0.7	0.7	0.7	0.7
United States	0.8	0.8	0.8	0.7	0.6	0.7
Uzbekistan	0.3	0.4	0.4	0.4	0.4	0.5
Mexico	0.4	0.4	0.4	0.4	0.3	0.4
Thailand	0.3	0.3	0.3	0.3	0.3	0.2
South Korea	0.3	0.3	0.3	0.2	0.3	0.2
Rest of World	2.2	2.1	2.0	1.9	2.0	1.9
Africa Franc Zone	0.0	0.0	0.0	0.0	0.0	0.0
EU-27	0.2	0.2	0.2	0.2	0.2	0.2
World total	24.0	24.4	24.5	25.0	24.4	26.3

culture between organic cotton agriculture, high levels of agrochemicals are used in the production of conventional cotton, chemicals used in the production of cotton pollutes the air and waters, decrease biodiversity, and destroyed of ecosystem equilibrium. In addition, the problem caused of conventional cotton agriculture is worse in developing countries with uninformed consumers, lack of stable institutions and property rights. Furthermore, thousands of farmers die because of the use of chemicals [5, 6].

In this study, organic agriculture, organic cotton agriculture, comparison conventional cotton agriculture with organic cotton agriculture, environmental impacts of organic cotton agriculture, and use of organic cotton products were investigated.

2   Organic Agriculture

In the world, agriculture industry has shown rapid change with the industrialization. As a result of rapid industrialization, uses of machines and chemical fertilizers have

Table 4 Word cotton imports

Word cotton imports						
Million metric tons	2013/14	2014/15	2015/16	2016/17	2017/18 February	2017/18 March
Bangladesh	1.2	1.3	1.4	1.5	1.4	1.6
Vietnam	0.7	0.9	1.0	1.2	1.0	1.5
China	3.1	1.8	1.0	1.1	1.0	1.1
Turkey	0.9	0.8	0.9	0.8	0.8	0.8
Indonesia	0.7	0.7	0.6	0.7	0.6	0.8
Pakistan	0.3	0.2	0.7	0.5	0.5	0.6
India	0.1	0.3	0.2	0.6	0.4	0.4
Thailand	0.3	0.3	0.3	0.3	0.3	0.2
South Korea	0.3	0.3	0.3	0.2	0.3	0.2
Mexico	0.2	0.2	0.2	0.2	0.3	0.2
Malaysia	0.1	0.1	0.1	0.1	0.2	0.1
Taiwan	0.2	0.2	0.2	0.1	0.1	0.1
Egypt	0.1	0.1	0.1	0.1	0.1	0.1
Rest of World	0.9	0.8	0.8	0.7	0.8	0.7
Africa Franc Zone	0.0	0.0	0.0	0.0	0.0	0.0
EU-27	0.2	0.2	0.2	0.2	0.2	0.2
World total	9.0	7.9	7.7	8.2	7.7	8.5

become widespread. Furthermore, increasing productivity in industry and technology causes to decrease production prices. When the negative effects of the methods used in agriculture are evaluated, innovative agricultural methods should be used in the agriculture industry.

At the present time, protection of environment and community health care has increased day by day. With the heavy reliance on chemical fertilizers and pesticides which pose a potential risk to human and animal health as well as to the environment, conventional agriculture disrupts natural balance, contaminating food chains and presenting threats to species and ecosystems.

Increase of environmental pollution has made organic agriculture which is eco-friendly production a current issue. Organic agriculture can be described as a system which shows regard to potential environmental and social impacts by ruling out the use of synthetic fertilizer, pesticides, veterinary drugs, genetically modified seeds, and additives. Organic agriculture provides to enhance agro-ecosystem health, biodiversity, biological cycles and soil biological activity. In this system that can be called as ecological agriculture, first aim is quality and the second aim is high efficiency [7].

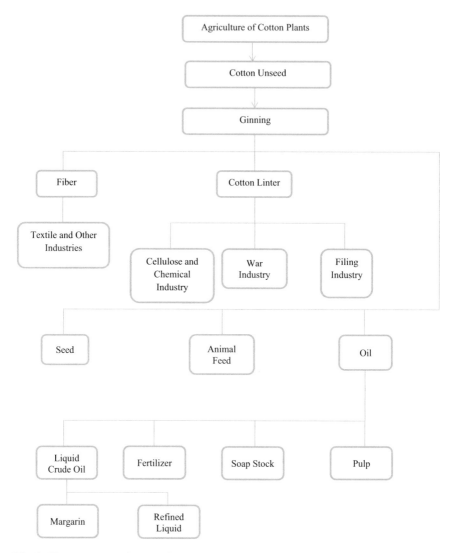

Fig. 1 The usage areas of cotton plants

History of organic agriculture dates back 20th century. In Europe, it began to apply in 1910s years, controlled production became widespread in 1930s years. In addition, it is called as "Green Revolution" which did not solve starvation problem of the whole world was deemed with discovering of pesticides and synthetic fertilizers. However, developed countries realized that pesticides and synthetic fertilizers affected of human health and ecological balance negatively in 1970s years and studies of organic agriculture was started. In 1972, International Federation of Organic Agriculture Movements (IFAOM) was established in Germany. This federation aims to gather all ecologic agriculture studies in the world under one roof, prepare of the necessity standard and regulations, transfer of the all development to producers and members [8].

Organic agriculture is one of sustainable agriculture system. Sustainable agriculture can be defined as an approach aims to balance environmental, social and economic dimensions. Furthermore, sustainable agriculture aims protection of efficiency, decrease in damage to the environment, to maintain the liveness of economy, and increase in life quality of farmers. The concept of sustainable agriculture is suggested to solve the problem of industrial agriculture. The basis of sustainable agriculture is that the sources of agriculture production are not unlimited and continuous development is not provided by disturbing the natural balance [9].

Sustainability is very significant for agriculture and it is related with ecosystem integrity, social well-being, good governance, and economic resilience. Furthermore, it has been equated to ecological strength in order to guarantee the continued provision of goods and services to present and future generations. Organic agriculture is described by the Codex Alimentarius Commission as an integral production management system which keeps away from use of pesticides, chemical fertilizers, genetically-modified organisms, lower soil, water and air pollution, optimizes human health and animal health, and efficiency of plants [10].

In organic agriculture, because of improvement of production efficiency, adaption of local conditions is necessary. The main organic strategies includes rotations, diversification and integration of crop, use of local varieties, use of biological pest control, and promotion of symbiotic nitrogen fixation and biomass recycling [10].

Organic agriculture system leads to a great amount of positive impacts such as enhanced soil efficiency, productivity, and better soil moisture retention, less water pollution, leaching of nitrate, decrease erosion, and higher soil carbon sequestration rates. Soil efficiency refers to combined use of independent and interdependent elements associated with potential of nutrient uptake, availability of soil nutrients and soil quality, which are indicators of increasing crop yields in a wide range of soil types. Site examination and excavating a soil pit which helps identify the properties of the soil profile are required for the evaluation of soil efficiency. Researches show that organic agriculture soil stocks organic carbon which is 3.5 metric tons per hectare higher than conventional agriculture soil. Furthermore, organic agriculture systems separate 450 kg more atmospheric carbon per hectare [10].

Compared with conventional agriculture, organic agriculture is more eco-friendly method that is not dependent on use of pesticides, herbicides and inorganic nutrient in the production of crops. As a result of many researches, organic agriculture prevents higher carbon storage of crops and higher levels of pesticides in water systems [11]. Moreover, organic agriculture provides to enhance bio-diversity in the agricultural nature, such as carabid beetles, vascular plants and birds. This production method is very important because conventional agriculture has resulted in a loss of bio-diversity in the agricultural nature. For example, the application of herbicides in conventional agriculture system causes to diminish weed abundances. At the same time, the use of pesticides decreases the predators which feed of pest insects [11].

Table 5 The differences of conventional and organic agriculture [21]

Conventional agriculture	Organic agriculture
Land is not detoxified	Land is changelessly detoxified
Synthetic fertilizers are used	Organic fertilizers are used
Synthetic/chemical pesticides and insecticides are used	Organic/herbal pesticides and insecticides are used Bio-controlling of insects is carried out
Seeds are treated with fungicides and insecticides	Seeds are not treated with fungicides and insecticides
Mono crop culture is applied	Crop rotation is applied
Genetically modified organism are used	Genetically modified organism are not used
Soil cannot keep water, lacks of organic materials and a lot of irrigational water is wasted	Soil has more water retention capacity owing to presence of organic material
Undesirable weeds are destroyed using herbicides	Undesirable weeds are manually or mechanically removed
Due to using chemicals, it can fatal effect on many friendly animals and insects such as owls, snakes, earthworms, frogs etc.	It is not effect bio-diversity and animal or insect habitat. Furthermore, it encourages their dwelling
Adversely affects ecological balance	Not adversely affects ecological balance
Defoliation occurs due to chemicals	Defoliation occurs seasonal

During the past 20 years, organic agriculture has enhanced in the many parts of the world. Because of organic agriculture future and position, many issues should be determined in the future. The first important issue is whether it can compete economically with conventional agriculture. This depends on cost of organic agriculture, quality of crops and demand of consumers. The second significant issue is climate conditions and environmental structure of soil. The climate conditions and soil structure are the most important matter effecting quality of crops. Furthermore, climate conditions and soil structure effects the type of crops. The interplay between the type of agriculture and biodiversity is also of prime importance. Unlike conventional agriculture, organic agriculture is likely to require more soil, thereby leading to lesser areas of natural and semi-natural ecosystems while maintaining higher levels of landscape biodiversity [12].

Conventional agriculture is known as industrial agriculture which includes the use of synthetic fertilizers, several of chemical, pesticides, herbicides, heavy irrigation, intensive tillage, and genetically modified organism. For these reason, conventional agriculture needs high resources and energy. However, this farming system is high productive [13]. There are many differences between conventional and organic agriculture. The differences of conventional and organic agriculture are given in Table 5.

Organic agriculture has increased rapidly all over the world each passing day. According to the 2017 data, 50.9 million hectares of organic agriculture land is and grows to more than 80 billion US Dollars in world. Demands of consumers have risen

Table 6 The data of organic agriculture in 2015–2016 years

Indicator	World	Top countries
Countries with organic activities	In 2015: 179 countries	Brunei Darussalam, Cape Verde, Hong Kong, Kuwait, Monaco, Sierra Leone, Somalia
Organic agricultural land	In 2015: 50.9 million hectares	Australia (22.7 million hectares) Argentina (3.1 million hectares) United States (2 million hectares)
Organic share of total agricultural land	In 2015: 1.1%	Liechtenstein (30.2%) Austria (21.3%) Sweden (16.9%)
Wild collection and further non-agricultural areas	In 2015: 39.7 million hectares	Finland (12.2 million hectares) Zambia (6.6 million hectares) India (3.7 million hectares)
Producers	In 2015: 2.4 million producers	India Ethiopia Mexico
Organic market	In 2015: 81.6 billion US dollars	US (39.7 billion US dollars) Germany (9.5 billion US dollars) France (6.1 billion US dollars)
Per capita consumption	In 2015: 11.1 dollars	Switzerland (291 US dollars) Denmark (212 US dollars) Sweden (196 US dollars)
Numbers of countries with organic regulations	In 2016: 87 countries	

and this increasing reflected in the market growth of 11% in the United States which is the largest organic market in the world. Number of organic farmers increased, more land was certified organic, and many of countries reported organic farming activities. The data of organic agriculture in 2015–2016 years were given in Table 6 [14].

3   Organic Cotton Agriculture

For a long time, in order to increase of products and control of pests, conventional pesticides and fertilizers have been used in cotton agriculture. In the world, pesticides used in cotton production are about 25% of pesticides used around the all agriculture. According to the Pesticide Action Network records, it has been attracted notice that costs invested in agrochemicals are very high and these agrochemicals are classified as "highly dangerous" by World Health Organization. These chemicals are harmful for environment, health, and life of employees. Every year, many employees who work in cotton farms are hospitalized as a result of contact with cotton plant agrochemicals. Moreover, pesticides are defined as a major global killer. In the world, nearly 1000 people die every day from acute pesticide poisoning and many more suffer from chronic ill health, such as lung cancers and leukemia, neurological diseases and reproductive problems including infertility, miscarriage and birth defects [15]. On the other hand, consequence of sustainable production methods and organic consumption has increased each passing day. Furthermore, efficiency of soil is protected with organic agriculture [16].

Global cotton production has improved since the 1930s. In cotton agriculture, high quantity toxic pesticides are used on the cotton crops [17]. Without the use of chemical fertilizers and pesticides, organic agriculture is an agricultural production system each stage of which is certified and controlled from production to consumption, which includes farming only with the ingredients permitted in accordance with the relevant regulations. In the present time, bio pesticides are used in the organic cotton agriculture and *Basillu Thuringiensis* is commonly used bio pesticide [18]. Since 1990s, textile and ready wear products have been involved in organic products market which improves with food products at the beginning. In the world, organic cotton production have been begun to become widespread since 1990s, especially production and usage of organic cotton has increased too in United States in 1995s. For these years, United States, Turkey, Argentina, Australia, Brazil, Benin, Egypt, Greece, India, Nicaragua, Paraguay, Peru, Tanzania, Uganda, Senegal and Mozambique are the important organic cotton producing countries [19].

As soon as reliable production figures can not be reached, United States and Turkey are the biggest countries of organic cotton producers according to Organic Trade Association (OTA) statistics [8]. International Federation of Organic Agriculture Movements describes organic agriculture according to four principles:

- Principle of health: Organic agriculture has not to damage human health, animal health and ecology. In addition, it has to enhance human, animal and plant health.
- Principle of ecology: Ecological systems and cycles of life are based on the organic agriculture.
- Principle of justice: Organic agriculture has to be established to provide justice for the agricultural environment and living facilities.
- Principle of care: Organic agriculture has to be a precaution to protect of current and future generation health [8].

The most important issue for cotton producer is economy in cotton production. For economical comparison of traditional cotton production with organic cotton production the following operations are required;

- The production stages should be determine for organic and conventional cotton agriculture separately,
- The costs should be determine for organic and conventional cotton agriculture,
- The unit area costs should be determine,
- The unit area revenue should be estimated,

Table 7 Comparison of conventional and organic cotton agriculture

Process	Conventional agriculture	Organic agriculture
Seed preparation	Seeds are treated with insecticides and fungicides	Seeds are not treated with any chemicals
Soil-water application	– Fertilizer is applied in the production – Water intensive – Soil loss can become due to water intensive	– Soil structure is conserved – Organic ingredients increase in soil – Unit area water is benefited further
Weed control	– Using herbicide before germinating – Continue using herbicide after germinating	– Using physical process for weed control – Weed control is carried out with weeding
Insect control	– Insecticides are used concentratedly – The insecticides used in insect control are high level toxic effect and some of them are cancerogenic – The insecticides are harmful for bio-diversity and ecology	– Insects which are harmful for cotton are eaten by other insects – Using of bait plant
Harvest	– Before harvest, drop leaves are performed with using toxic chemical	– Drop leaves occurs seasonal frosts
Production finance	– Finance can be carried out with borrowing	– Limiting finances

– Investment analysis should be carried out for short and long term.

Compared with conventional cotton agriculture and organic cotton agriculture, there are many significant differences. These differences are given in Table 7.

Compared with conventional and organic cotton agriculture, there are differences the use of soil and water. The amount of soil and water consumptions was given in Tables 8 and 9 [3].

Cotton is a very water-intensive crop; it is estimated that cotton growing results in 1–6% of the world's total freshwater withdrawal. In order to produce 1 kg of cotton lint, 10,000–17,000 L water is required [3].

In organic cotton agriculture, feeding of cotton plants by organism is essential. It is made real that sources of organic nutrients are given to soil and these nutrients are transformed into nutrients for plants with saturating in the soil. In order to develop of soil efficiency and biological activity, the application of product rotation is the most recommended method. Product rotation method has many advantages. This method breaks down the cycle of insect damage and plant diseases. With the selection of suitable rotation products and planting temporization, the weed problems can

Table 8 Conventional and organic cotton agriculture in the world

	Conventional cotton			Organic cotton		
	Production (m ton)	Cotton % among fibers	Cotton cultivated area (mha)	Production (ton)	Organic cotton %	Organic cotton cultivated area (ha)
1960	10.113	67.5	30–36			
1999–2000	20.2	37.5		7545	0.04	
2000–2001	18.869	38.1		6480	0.03	
2001–2002	21.281	38.4		18,000	0.08	
2003–2004	21.135	36.9		25,394	0.10	
2005–2006	26.532	38.6	30–36	37,799	0.14	
2006–2007	26.751		34–36	57,931	0.20	
2011–2012	27.100		35–52	139,000	0.50	317,000
2013–2014	25.700			116,974	0.40	220,765
2015–2016	21.87		30–49			

Table 9 Water consumption in organic cotton agriculture

Water consumption of organic cotton	
Country	Water consumption (m^3/ha)
India	2950–4100
USA	1500–4500
Israel	1300–2800

be reduced. Product rotation can affect the soil efficiency a considerable extent. Including legumes used as animal feed to rotation leads to nitrogen for organic cotton. In product rotation method, the use of plant covering the soil surface prohibits erosion. Furthermore, soil thrives in terms of nitrogen with the mingling of these plants in the soil. Growing rapidly and firmly and covering the soil surface, plants have been effective as smother or allelopathic cover in suppressing weeds [18].

In conventional cotton, there are many factors effecting human health. These parameters can be listed as;

– Residue of pesticides: Pesticide is described as a chemical used to remove any living. Pesticide is a chemical general name which is used for insects, fungi, rodentia, and weeds. Pesticides consist of several subgroups as insecticides (for insects), fungicides (for fungi), herbicides (for plant), rodentia (for rodents), repellents (for insect repellent), acaracides (for mites), attractants (for insects), nematocides (for worms), arboracides (for brush), and algicides (for algaes).

– Heavy metal ions: The penetration of heavy metal ions into textile products can occur in various forms. Defoliants or pesticides can lead to penetration of heavy metal ions to cotton fibers. The heavy metal ions such as chrome, mercury, lead, arsenic, cadmium, cobalt, nickel, nickel salts, copper and zinc [16].

The fruitfulness of cotton is limited because of the following external factors: scale of production, level of research support, local ginning capacity, access to quality seed, access to irrigation, access to timely inputs, production costs, price of cotton seed, access to credit, timely payment for the crop and education of farmers. For conventional cotton production, the biggest sustainability challenge is the use of agrochemicals which are known as negative effects on human health and potential harm to the environment. Furthermore, conventional cotton agriculture is not suitable for smallholder farmers. Organic agriculture is a favorable alternative to such farmers with potential advantages of lower expanses for farm inputs and healthier soils [20].

The advantages of organic cotton can be listed as,

– Production process does not include use of chemicals,
– Compared with conventional cotton, organic cotton has better quality,
– Commercially beneficial,
– Health-friendly and environmentally-friendly,
– Ensures better health for farmers and their families,
– Reduced pollution which pesticides causes,
– Prevents water contamination and protect biodiversity,
– Being skin-friendly fiber [21].

People do not have enough information about organic cotton. Some people believe that conventional cotton has better quality than organic cotton, however, this belief is not correct. Organic cotton is better quality than conventional cotton. Because the amount of production is more, various farmers prefer to conventional cotton production. However, it was true even a few years ago. Cotton plant is inclined to infestation and fungal infections and cotton farmers cultivate more cotton products with using lots of synthetic insecticides, pesticides, fungicides, and fertilizers. Furthermore, advanced techniques of organic cotton farming were limited to investigate for few decades ago, but this changed at the present time. Egypt, a major cotton producer, developed an advanced organic cotton farming technique which reduced usage of synthetic fertilizers and pesticides by more than 90% and achieved a growth of 30% in production.

When taking into consideration of organic cotton future, it is been estimated that the demands of organic cotton will increase. A growing number of consumers prefer organic cotton products such as personal care products, accessories and clothes of baby and children. For this reason, organic cotton market is a special market [16].

In order to commercialize of organic cotton, it requires a certification including farm production applications. The certification should be continued in the production process. At every phase of production, materials suitable for organic specifications (dyes, bleaches, etc.) should be used. Only the clothes produced on these processes can be sold as "organic". In order to distinguish organic product, organic products should include suitable labels. At the same time, these labels serve to understand easily explanation of human health and production ecology. At the present time, there are several of products having eco-textile label. Some part of these labels is associated with human health, while other labels focus eco-friendly production [16].

Organic cotton agriculture is more risky than conventional agriculture with regards to producers. A farmer who produces cotton with conventional methods does not change the production method following year. For organic agriculture, a transition period of at least 3 years is required. Furthermore, organic fibers are more expensive than conventional fibers due to high production costs. However, there is an increased demand of organic fibers despite the increased costs [19].

Guerena and Sullivan investigated the comparing of conventional cotton production costs between organic cotton production costs. The results show that the costs of organic cotton production are more than 50% costs of conventional cotton production. Furthermore, compared with conventional cotton products between organic cotton products, it was reported that there are not significant differences in terms of quality. Nevertheless, because of the higher costs of organic cotton production and the lower efficiency of organic cotton production, it was reported that price premium must be paid for organic cotton producers in order to prevent lose money. In organic cotton production, the factors caused to differences between conventional cotton production are given in below;

– Fertilizing materials (organic fertilizers, compost),
– Costs of struggling with weed by mechanically,
– Insect and disease-fighting materials suitable for organic agriculture (compost tea, insects),
– Costs of workforce for pulling of weeds by hands,
– Costs of taking of organic cotton production certification [22].

Although the organic cotton agriculture costs of harvests, cultivation, transport and stocking is the same as conventional cotton agriculture, organic cotton agriculture requires higher workforce. In organic agriculture, protection methods of soil and struggle with insects and weeds are more expensive than the use of chemical methods. In addition, organic cotton producers have to observe certifications and these certifications brings some costs [19].

After the production of conventional cotton, it can be sold cooperatives, individual merchants and private sector. First, unseed cotton is ginned and this process causes to occur many cotton seeds. Then, these cotton seeds are processed in oil industry and cotton oil is obtained. After the obtaining of cotton oil, pulp occurs and these pulps are used as raw material in the pulp industry. Furthermore, after the ginning process, cotton fibers are obtained and these fibers are processed in yarn factories, weaving factories and ready to wear factories.

In comparison to conventional cotton market with organic cotton market, the structure of organic cotton market is different from the structure of conventional cotton market. In commercialize of organic cotton; producer, certification firm and purchaser come to terms on organic agriculture. Farms are controlled by certification firms and acceptance of appeal is evaluated. In case of the acceptance of appeal, these producers are a member of Ecological Farming Association. After this, producers can sell organic product to firms or staple freely [19]. Producers take crops to ginning factory and cotton crops are ginned. After, cotton fibers are pressed and cotton bales are obtained. These bales can export or be sent to yarn factory in order to produce

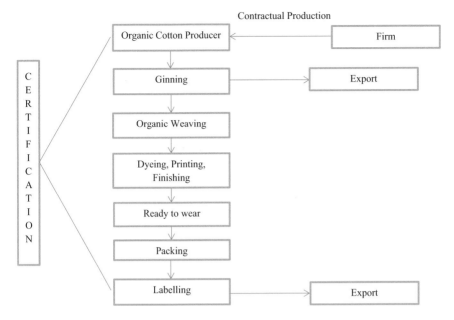

Fig. 2 The marketing structure of organic production [28]

cotton yarns. After this stage, organic cotton yarns can be sent to weaving, knitting, dyeing, finishing or ready to wear factory. The organic cotton production process is given in Fig. 2.

4 Use of Organic Cotton Products

Organic agriculture is one of the most important alternative production system protected of soil benefits, environment, health, farming communities economy. In the world market, demand of organic products has been raised each passing day. In order to become widespread of organic agriculture in the production of all crops, farmers should be raised awareness and encouraged with economically [23].

Fast growth in organic markets has caused to increase global demand for organic products. In developing countries, organic sector which shows significant economic growth has proceed towards export. In some of the developing countries, there have been powerful calls for promoting to organic agriculture and some developing countries have begun to apply of organic policies [23]. These developments induced to be a considerable opportunity for small and medium farmers in these countries.

In organic cotton market, so as cotton products can be sold as "organic" label, it requires certification of the field production practices. Certification must be given upwards the manufacturing process, from ginner, yarn spinner, and clothing. Each step of the process (dyeing, bleaching, etc.) must be carried out meeting organic specifications. The products which are not already on the National Organic Program's approved list have to undergo a long process to acquire approval [22].

Organic cotton can purely be called as skin-friendly, restful, and harmless natural fiber. Compared with conventional cotton, organic cotton is not irritating to newborn skin. For this reason, organic cotton is the ideal material for protecting and cleaning newborn babies, producing for clothes, towels, cleaning wounds, and every textile material for newborn babies. Furthermore, organic cotton can be used in surgeries which contamination from any source can be fatal. Organic cotton seeds have been used in livestock farming and oil industry [21].

Organic cotton clothing has some of benefits such as better health for consumers and farmers, environmental protection, and cost savings for farmers. Due to these benefits, organic cotton clothing industry has been blown out in the past few years. Especially, parents are increasingly purchasing organic cotton baby clothes, baby diapers, and baby blankets. Several consumers which are women to men, young to old have begun to prefer various products made from organic cotton such as jeans, pajamas, t-shirts, towels, beddings, mattresses, pillows, blankets, sheets, duvet covers, socks, bath robes, underwear, bags, and other clothes made of organic cotton fabric and yarn. Producers of organic cotton clothing strongly advertise that organic cotton clothes are suitable especially for babies. Several consumers and parents also think in this manner. It is known that organic cotton clothing is softer than conventional cotton clothing. In addition, because any chemicals are not used in production of organic cotton, it does not cause to allergies for users. Organic cotton causes also to reduce respiratory problems and smells pleasant [24]. The advantages of organic cotton clothes are listed as:

– Since chemicals are not used in the production, environmental damage is reduced.
– In the production, bleaching process is not used.
– In some cases, use of chemicals is not necessity to dye of organic cotton yarns or fabrics.
– Because any chemicals are not used in the manufacturing of organic cotton products, allergic disorders disappear from consumers with sensitive skin (especially babies).
– With the fiber production, since cotton oil which is produced from cotton seeds is used in food industry and cotton seed pulp is used as animal feed, organic cotton agriculture is important in terms of healthy eating of people.
– Because organic cotton agriculture requires a certain amount of knowledge and techniques, it plays an essential role in increase of knowledge and culture levels of farmers.
– All the products of the producer is guaranteed to be purchased with contract production.
– Revenue of producers increase (average 10%) as well as other organic agriculture products.

Besides these benefits of organic cotton clothes, the high price is the most important disadvantage of organic cotton clothes. Cost of organic agriculture is about 50%

more than conventional cotton agriculture. For this reason, high cost of organic cotton agriculture causes to increase in the costs of products. The prices of clothes produced with 100% organic cotton is 20–50% more expensive than the prices of clothes produced with conventional cotton [8].

Since the past twenty years, organic cotton market has showed rapid growth. However, the amount of organic production does not meet the demands. In newborn baby textile market, parents have preferred to buy organic cotton products. Compared with conventional cotton products between organic cotton products, prices of organic cotton products are higher than conventional cotton products. The disadvantage of organic cotton products is the high prices.

Rather than the commercial gain, organic cotton product is health-friendly and environmental friendly, unlike conventional cotton product.

Although the high prices of cotton and cotton products, proportion of land separated for organic cotton and demand of consumer have continued to rise. However, for consumers to adopt organic textile products, manufacturers should be invested in organic cotton products further more. In order to decrease in prices of organic cotton products, enhancing of target oriented marketing, certifying and increase in variety of organic cotton products.

Early on, organic cotton was used to manufacture of new baby born clothes. However, at the present time, organic cotton is used:

– Personal care products (medical products, make-up cleaning products, powder puffs, cotton buds),
– Babies and children clothes and accessories,
– Ready to wear products for people having allergic habit of body,
– Towels, bathrobes, blankets and beddings [8].

5 Environmental Impacts of Organic Cotton Agriculture

"Organic agriculture" and "sustainable agriculture" new terms came to exist with the damaging of agriculture to human health and environment. Organic agriculture causes to decrease in intensive inputs, protect of soil and water. For these reasons, organic agriculture came to the fore in developed countries. With the use of intensive inputs in the agriculture, contamination of water and soil increases, residual of pesticides occurs on the crops, and plant mites gain durability to chemicals. The aims of organic agriculture can be explained as improvement of efficiency, decrease in damage to the environment, keeping the economy alive over the long and short run, and increase in life standards of farmers.

First serious investigations about organic cotton agriculture began in Southeastern Anatolia, Egypt, and southern regions of the United States in 1980s. Organic cotton agriculture is aimed to solve environmental pollutions which originates of conventional cotton agriculture [16].

Although cotton is cultivated on only 3% of the total agricultural area, 25% of the insecticides and 10% of the pesticides are used in the cotton agriculture. Cotton crops have copes with numerous pests and diseases over the years such as bacterial blight, black root rot, sting nematode, lint degradation, lint contamination, cotton rust, blue diseases, small leaf, leaf roll, psylosis, terminal stunt, powdery mildew, verticillium wilt, and phyllody [24]. The extreme use of chemicals in conventional cotton agriculture has caused to important environmental pollution. Due to spread the environmental pollution and increase in the demands of "eco-friendly products", organic cotton has been an alternative to conventional cotton which is cultivated with the use of pesticides and other toxic materials.

On the market, cotton sold with organic cotton label must be cultivated according to guidelines of organic crop production. Soil efficiency exercises which meet organic certification standards contain cover cropping, crop rotation, addition of animal manure, and use of natural occurring rock powders. Weed control can be provided with flame weeding, cultural practices and combination of cultivation. Cotton plants are attacked by a variety of insects in growing stage. In order to tackle with insects, trap cropping, strip cropping and managing border vegetation can be applied in cotton farms. Furthermore, bacteria, viruses, and fungal insects pathogens can be used in insects control [22].

Increase in the selling of organic cotton products, label play an essential role. Consumers continuously collect information about evaluating and choosing products. Manufacturers can obtain which products are preferred to buy by consumers with investigated the labels of products. In other words, due to involving product information, the labels of products affected to be preferred by consumers. Researchers attract notice the importance of labels owing to making choice of consumers. Labels are found to supply consumers with useful information on making decision. Recently, with increase in environmental consciousness, manufacturers have focused on eco-friendly production and they have to prove to consumers with labels. Researches show that about 60% of consumers actually read labels when shopping for clothing and the label information affects their decision making. Especially, organic cotton clothing producers improved successful marketing strategies with using labels in order to increase the consumption of organic cotton products [25].

6   Conclusion

Since pre-historic times, cotton fibers have been preferred by most people and these fibers is the most used fiber at the present time. Cotton fibers are cultivated on 34 million hectares globally (7% of the world's arable land). Statistics indicates that 70 million tons of cotton is produced annually and 25 million tons used by textile industry in order to manufacture fabrics, yarns, clothing, medicine textiles, and etc. These data show that cotton fibers are significant for textile industry [26].

In the world, because agricultural products can not feed the growing population, researchers have sought for new agricultural methods. In order to increase the quality of products to be obtained from the unit side, use of synthetic agriculture materials such as chemical fertilizers, hormones, and agricultural pesticides made widespread. The agriculture methods have caused to distribution of ecological balance, chemical wastes in crops threaten human health, the degradation of plant and animal health, and increase in cost of production. Producers and consumers have begun to not prefer agricultural products which do not effect human health and ecology negatively to remove these problems, with these aims, organic agriculture methods have occurred [27].

In 1970s years, organic cotton agriculture attracted attention as a result of increase in environmental consciousness. In conventional cotton agriculture, too much chemicals are used without considered of natural life cycle. Consequently, farmer health and ecology are influenced adverse.

Recently, organic cotton market has enhanced rapidly in the world. But, the demand of organic cotton production is insufficient. A lot of parents have preferred to buy organic cotton products for newborn baby, although, prices of organic cotton products are higher than conventional cotton products.

References

1. Dayioglu H, Karakas HC (2007) Elyaf Bilgisi. Istanbul, Turkey
2. https://arastirma.tarim.gov.tr/tepge/belgeler/TARIM%20%C3%9CR%C3%9CNLER%C4%B0%20P%C4%B0YASA%20RAPORLARI%20KLAS%C3%96R%C3%9C/Pamuk%20Tar%C4%B1m%20%C3%9Cr%C3%BCnleri%20Piyasalar%C4%B1ndaki%20Geli%C5%9Fmeler.pdf. 17 Mar 2018
3. Uygur A (2017) The future of organic fibers. Eur J Sustain Develop Res 2(1):164–172
4. http://www.cottoninc.com/corporate/Market-Data/MonthlyEconomicLetter/pdfs/English-pdf-charts-and-tables/World-Cotton-Consumption-Metric-Tons.pdf. 12 Mar 2018
5. https://organiccottonplus.com/pages/learning-center. 15 Mar 2018
6. http://www.wiki-zero.com/index.php?q=aHR0cHM6Ly9lbi53aWtpcGVkaWEub3JnL3dpa2kvT3JnYW5pY19jb3R0b24. 12 Mar 2018
7. http://www.fao.org/organicag/oa-faq/oa-faq1/en/. 04.03.2018
8. Koramaz NE (2011) Investigation of dyeing properties of organic cotton. M.Sc. Thesis, Erciyes University, Graduate School of Natural and Applied Science, Kayseri, Turkey
9. Demiryürek K (2011) The concept of organic agriculture and current status of in the world and Turkey. J Agric Fac Gaziosmanpasa Univ 28(1):27–36
10. Scialabba NH (2013) Organic agriculture's contribution to sustainability. Plant Manage Netw. http://www.fao.org/docrep/018/aq537e/aq537e.pdf
11. Bengtsson J, Ahnström J, Weibull AC (2005) The effects of organic agriculture on biodiversity and abundance: a meta-analysis. J Appl Ecol 42(2):261–269
12. De Ponti T, Rijk B, Van Ittersum MK (2012) The crop yield gap between organic and conventional agriculture. Agr Syst 108:1–9
13. http://www.appropedia.org/Conventional_farming. 08 Mar 2018
14. Seufert V, Ramankutty N, Foley JA (2012) Comparing the yields of organic and conventional agriculture. Nature 485(7397):229
15. http://www.pan-uk.org/cotton/. 06 Apr 2018
16. Kasapoglu O (2007) Organic cotton and organic cotton yarn cost analysis. M.Sc. Thesis, Istanbul Technical University, Graduate School of Natural and Applied Science, Istanbul, Turkey
17. http://el.doccentre.info/eldoc1/g74a/01dec01LEISA6.pdf. 06 Mar 2018

18. Gurkan S (2012) Finishing methods of organic cotton knitted fabrics by ultrasound technology. M.Sc. Thesis, Istanbul Technical University, Graduate School of Natural and Applied Science, Istanbul, Turkey
19. Keskin U (2007) The production and economics of organic cotton in turkey and in the world. M.Sc. Thesis Cukurova University, Graduate School of Natural and Applied Science, Adana, Turkey
20. Riar A, Mandloi LS, Poswal RS et al (2017) A diagnosis of biophysical and socio-economic factors influencing farmers' choice to adopt organic or conventional farming systems for cotton production. Front Plant Sci 8:128
21. https://www.organicfacts.net/organic-cotton.html. 11 Mar 2018
22. Guerena M, Sullivan P (2003) Organic cotton production. Appropriate Technology Transfer for Rural Areas (ATTRA), https://attra.ncat.org/attrapub/PDF/cotton.pdf
23. Kolanu TR, Kumar S (2007) Greening agriculture in india: an overview of opportunities & constraints. Food Agriculture Organization of the United Nations (FAO), Hyderabad
24. https://www.organicfacts.net/organic-cotton-clothing.html. 10 Apr 2018
25. Oh K, Abraham L (2016) Effect of knowledge on decision making in the context of organic cotton clothing. Int J Consum Stud 40(1):66–74
26. Garcia S, Bernini DDS, Nääs IDA et al (2015) Colored and agroecological cotton may be a sustainable solution for future textile industry. GEPROS, Gestão da Produção Operações e Sistemas (Online) 10(1):87–100
27. Yolcu N (2013) Organic agriculture and potential for employment creation of organic agriculture in Turkey. M.Sc. Thesis, Karadeniz Technical University, Graduate School of Natural and Applied Science, Trabzon, Turkey
28. Aksoy E, Dolekoglu T (2003) Dünyada ve Türkiye'de Organik Pamuk Üretim ve Ticareti. Türkiye VI. Pamuk, Tekstil ve Konfeksiyon Sempozyumu, pp 61–62
29. http://koop.gtb.gov.tr/data/5ad06c80ddee7dd8b423eb24/2017%20Pamuk%20Raporu.pdf. 28 July 2018

Printed in the United States
By Bookmasters